**BOTANICAL RESEARCH AND PRACTICES**

# HERBACEOUS PLANTS

# CULTIVATION METHODS, GRAZING AND ENVIRONMENTAL IMPACTS

# BOTANICAL RESEARCH AND PRACTICES

Additional books in this series can be found on Nova's website under the Series tab.

Additional e-books in this series can be found on Nova's website under the e-book tab.

BOTANICAL RESEARCH AND PRACTICES

# HERBACEOUS PLANTS

# CULTIVATION METHODS, GRAZING AND ENVIRONMENTAL IMPACTS

FLORIAN WALLNER
EDITOR

*New York*

Copyright © 2013 by Nova Science Publishers, Inc.

**All rights reserved.** No part of this book may be reproduced, stored in a retrieval system or transmitted in any form or by any means: electronic, electrostatic, magnetic, tape, mechanical photocopying, recording or otherwise without the written permission of the Publisher.

For permission to use material from this book please contact us:
Telephone 631-231-7269; Fax 631-231-8175
Web Site: http://www.novapublishers.com

### NOTICE TO THE READER

The Publisher has taken reasonable care in the preparation of this book, but makes no expressed or implied warranty of any kind and assumes no responsibility for any errors or omissions. No liability is assumed for incidental or consequential damages in connection with or arising out of information contained in this book. The Publisher shall not be liable for any special, consequential, or exemplary damages resulting, in whole or in part, from the readers' use of, or reliance upon, this material. Any parts of this book based on government reports are so indicated and copyright is claimed for those parts to the extent applicable to compilations of such works.

Independent verification should be sought for any data, advice or recommendations contained in this book. In addition, no responsibility is assumed by the publisher for any injury and/or damage to persons or property arising from any methods, products, instructions, ideas or otherwise contained in this publication.

This publication is designed to provide accurate and authoritative information with regard to the subject matter covered herein. It is sold with the clear understanding that the Publisher is not engaged in rendering legal or any other professional services. If legal or any other expert assistance is required, the services of a competent person should be sought. FROM A DECLARATION OF PARTICIPANTS JOINTLY ADOPTED BY A COMMITTEE OF THE AMERICAN BAR ASSOCIATION AND A COMMITTEE OF PUBLISHERS.

Additional color graphics may be available in the e-book version of this book.

**Library of Congress Cataloging-in-Publication Data**

Herbaceous plants : cultivation methods, grazing and environmental impacts / editor: Florian Wallner.
   p. cm.
 Includes index.
 ISBN 978-1-62618-729-0 (hardcover)
 1. Herbaceous plants. 2. Grazing. 3. Energy crops. I. Wallner, Florian.
 SB404.9.H46 2013     582.1'2--dc23
                                                     2013013157

*Published by Nova Science Publishers, Inc. † New York*

# Contents

| | | |
|---|---|---|
| **Preface** | | vii |
| **Chapter 1** | Exploring the Role of the Diversity of Herbaceous Feed Items for Shrubby Rangeland Management<br>*Laíse da Silveira Pontes, Cyril Agreil, Danièle Magda, Hervé Fritz and Pedro Gonzalez-Pech* | 1 |
| **Chapter 2** | Biosynthesis and Regulation of Tobacco Alkaloids<br>*Tsubasa Shoji and Takashi Hashimoto* | 37 |
| **Chapter 3** | Hybrid Lethality in Nicotiana: A Review with Special Attention to Interspecific Crosses between Species in Sect. Suaveolentes and N. Tabacum<br>*Takahiro Tezuka* | 69 |
| **Chapter 4** | Tolerance of Herbaceous Plants to Multiple Contaminations in an Industrial Barren Near a Nickel-Copper Smelter<br>*R. Kikuchi, T. T. Gorbacheva, M. V. Slukovskaya and L. A. Ivanova* | 95 |
| **Chapter 5** | Effects of Abiotic Factors on Herbaceous Plant Community Structure: A Case Study in Southeast Cameroon<br>*Jacob Willie, Eduardo de la Peña, Nikki Tagg and Luc Lens* | 113 |

| **Chapter 6** | Environmental Performance of Three Novel Opportunity Biofuels: Poplar, Brassica and Cassava during Fixed Bed Combustion<br>*Maryori Díaz-Ramírez, Christoffer Boman, Fernando Sebastián, Javier Royo, Shaojun Xiong and Dan Boström* | **133** |

**Index** **149**

# Preface

In this book, the authors present current research in the study of herbaceous plants and nicotiana. Topics include the role of diversity of herbaceous feed items for shrubby rangeland management; the biosynthesis and regulation of tobacco alkaloids; hybrid lethality in nicotiana; tolerance of herbaceous plants to multiple contaminations in an industrial barren near a nickel-copper smelter; effects of abiotic factors on herbaceous plant community structure; and the environmental performance of three novel opportunity biofuels: popular, brassica and cassava.

Chapter 1 – Biodiverse pasture ecosystems, such as rangelands, are now highly valued for their ecological, landscape and agronomic properties. The vegetation of these plant communities usually consists of a variety of herbaceous and shrubby species. Grazing and browsing by domestic herbivores has been proposed as an economic and efficient way to restore biodiversity and forage resource quality in herbaceous-shrub mosaics by, for example, avoiding shrub encroachment. Further, novel concepts in grazing science draw attention to the importance of maintaining functional heterogeneity in order to sustain food intake through behavioral adjustment. However, the challenge is now to define the heterogeneity, allowing integration of productive goals and the control of shrub dynamics in order to guide practices for rangeland management. Therefore, in this chapter, the authors aim to analyze the recent conceptual advances in the role of diverse vegetation on intake dynamics. The authors explore the foraging responses of ruminants faced with a diversity of herbaceous feed items and their effects on shrub (e.g. *Cytisus scoparius* Linck) consumption. The authors aim also to take into account the role of these feeding choices on intake dynamics at different time scales. Our approach is mainly based on experiments carried out in southern France on ewes grazing relatively small fenced paddocks for short periods of time in shrubby rangelands. Flock activities were always recorded through scan sampling method, and the ewes' diet selection was encoded as bite categories. For instance, the effects of three different herbaceous covers with different forage availability and quality on shrub consumption were compared, mainly by showing how the availability of "herbaceous bites" and selection affect the way that ewes integrate shrubs into their diet. The authors also identify the behavioral adjustment possibilities between the size and quality of herbaceous *vs.* shrubby bites into meals. The results provide insights into ways to manipulate diet selection in order to stimulate the use of a dominant shrub by ewes. They also highlight how grazing management of herbaceous physiognomy can lead to efficient shrub encroachment control. The authors discuss the usefulness of bite categories as functional feed indicators. Finally, operational implications are discussed in order to implement original approaches that take into account the functional interactions between ecological and technical processes. Adaptive management and its operational implementation are suggested as a promising approach.

Chapter 2 – In *Nicotiana* plants, nicotine and related pyridine alkaloids, such as nornicotine, anabasine and anatabine, are synthesized in underground roots and then translocated via the xylem to aerial parts, where they are mainly stored in vacuoles as defensive toxins against herbivorous insects. A series of structural genes involved in the synthesis and transport of these alkaloids have been isolated, mostly based on their homology or expression profiles, and shown to be expressed in distinct types of root cells. Nevertheless, enzymes catalyzing the late synthetic steps, including ring coupling, have remained elusive. Jasmonate signals, in response to insect hervibory, and cross-talking with auxin and ethylene, coordinately activate the nicotine pathway genes through a signaling cascade consisting of *Nicotiana* COI1, JAZs, and the bHLH transcription factor MYC2. Two genetic loci, *NIC1* and *NIC2*, mutant alleles of which have been used to breed tobacco cultivars with low nicotine content, specifically control multiple structural genes of the nicotine pathway. A group of closely related *ERF* transcription factor genes are clustered at the *NIC2* locus and deleted in the *nic2* mutant. Jasmonate-inducible *NIC2*-locus ERFs and MYC2 directly up-regulate the transcription of the nicotine pathway genes, recognizing specific *cis*-elements in the promoters of their downstream target genes.

Chapter 3 – Inviability of hybrids, often referred to as hybrid lethality, is a type of reproductive isolating mechanism. Hybrid lethality is observed in a number of plant species, including *Nicotiana* species, and can be an obstacle to the introduction of desirable genes into cultivated species by wide hybridization. In this chapter, the author reviews hybrid lethality in *Nicotiana* with special attention to interspecific crosses between species in *Nicotiana* sect. *Suaveolentes* and *N. tabacum*. Most wild species in sect. *Suaveolentes* (which consists of species restricted to Australasia and Africa) yield inviable hybrids after crosses with the cultivated species *N. tabacum* ($2n = 48$, SSTT). Genetic studies have revealed that hybrid lethality in *N. tabacum* × *N. debneyi* is caused by interaction between one or more genes on the Q chromosome of *N. tabacum* and a single dominant gene in *N. debneyi*. Gene(s) on the *N. tabacum* Q chromosome are also responsible for hybrid lethality in crosses involving most other *Suaveolentes* species. Most notably, genes from both S and T subgenomes of *N. tabacum* are responsible for hybrid lethality in *N. tabacum* × *N. occidentalis*. In addition, two species, *N. benthamiana* and *N. fragrans*, produced 100% viable hybrids from crosses with *N. tabacum*. These results provide a framework for discussing evolutionary processes leading to hybrid lethality in sect. *Suaveolentes*.

Chapter 4 – A better understanding of heavy metal sources, their accumulation in the soil and the effect of their presence in soil on plant systems seems to be particularly important in present-day research on risk assessments. It is necessary to evaluate plant tolerance when discussing this subject.

The most severe effects of metals on ecosystems are from local pollution in the Arctic/subarctic regions, and the Kola Peninsula (66–70°N and 28°30'–41°30'E) in Russia is one of the most seriously polluted regions: close by nickel-copper smelters, the deposition of metal pollutants has severely damaged the soil and ground vegetation, resulting in an industrial barren. During 2011–2012, a field test was performed near the smelter complex (67°51'N, 32°48'E). The applied method is based on cultivation of perennial grasses using hydroponics with vermiculite from a local deposit followed by rolled lawn placement on the metal-polluted sites. To avoid root system disturbance, the authors used an additional 5 cm-layer from local deposit carbonatites (lime-like materials). Original carbonatites show high initial nutritional status: bioavailable forms of 18 mg kg$^{-1}$ K, 123 g kg$^{-1}$ Ca, 1.8 g kg$^{-1}$ Mg and 89 mg kg$^{-1}$ P. Multiple pollution was observed during the field test: the precipitation amount of $SO_4^{2-}$ in the study field was over 5.57 times (4406,3 g ha$^{-1}$) greater than that in the background field, the Cu amount in the study field was over 645 times (572,5 g ha$^{-1}$) greater than that in the background field, and the Ni amount in the study field was over 824 times (685,3 g ha$^{-1}$) greater than that in the background field. The results obtained from leaf diagnostics also show that the monitored plants are tolerant to multiple stress (cf. monitored zone vs. background zone): N – 15942 mg kg$^{-1}$ vs. 11300 mg kg$^{-1}$; P – 608 mg kg$^{-1}$ vs. 1660 mg kg$^{-1}$; K – 17266 mg kg$^{-1}$ vs.12290 mg kg$^{-1}$; Ca – 5388 mg kg$^{-1}$ vs. 1700 mg kg$^{-1}$; Mg – 1947 mg kg$^{-1}$ vs. 650 mg kg$^{-1}$. The authors' observation is still continuing in order to study the influence of freezing and the nutrient loss rate.

Chapter 5 – Abiotic factors significantly influence the structure of plant communities, with the effects varying in both space and time. Herbaceous plants belonging to 15 families were monitored in 250 4-m² plots distributed in six habitat types in order to assess the effects of abiotic factors on the abundance of this resource. In each plot, the authors counted herb stems and determined the total number of species, the total number of normal stems and the total number of dwarf stems. In addition, they determined soil fertility and other environmental variables. Elevation and soil texture varied, but similar levels of chemical fertility were seen across different habitat types. Herb abundance varied within and between patches, reflecting changes in environmental conditions. Stem biomass was highest in light gaps, and decreased in late successional forests. Light seemed to be the most important factor influencing the abundance of herbs from Marantaceae and Zingiberaceae families only. Despite the hydromorphic nature of the soil in swamps, stem biomass did not exceed that of *terra firma* forests. At the temporal scale, rainfall did not seem to influence stem density as herbaceous plants were available year-round. These results suggest that light might limit the abundance of some herbaceous plants in the study site. However, a long-term investigation is needed to draw firm conclusions on the effects of abiotic factors on herbaceous plant communities in African rain forest.

Chapter 6 – In the last few decades several types of solid biofuels have been proposed as possible sources for heat generation because of growing concerns about environmental pollution, and future fossil fuel supply uncertainties. Among other biomass assortments, short rotation coppice and herbaceous species have been considered. An important aspect to be evaluated to enable a sustainable introduction of such novel fuels is related to their environmental performance during combustion. In this work, three fuel types; one herbaceous energy crop and one short rotation coppice (both cultivated and pelletized in Spain), together with one agricultural residue (cultivated in China) have been assessed in terms of their emission levels of gases (CO and $NO_X$) and particulate matter. The experiments showed that combustion of the fuels was attained under an acceptable level of CO emissions.

However, concentration of $NO_X$ was rather high, but perhaps more important, a considerably high formation of fine particle emissions was observed. Consequently, the incorporation of primary or secondary particle precipitating reduction measures might be needed. In addition, the high ash content in these fuels can severely deteriorate the combustion performance and reliability. Thus, specially designed burners/grate units are therefore needed if a utilization of these fuels in small and medium scale combustion systems seeks to be feasible. Although the applicability of introducing this kind of biofuels to the residential heating sector perhaps seems to be rather limited, it should not always be rejected. Nevertheless, technology improvements would have to be considered to manage the current limitations.

In: Herbaceous Plants
Editor: Florian Wallner

ISBN: 978-1-62618-729-0
© 2013 Nova Science Publishers, Inc.

*Chapter 1*

# Exploring the Role of the Diversity of Herbaceous Feed Items for Shrubby Rangeland Management

*Laíse da Silveira Pontes[1], Cyril Agreil[2,3]\*, Danièle Magda[4], Hervé Fritz[5] and Pedro Gonzalez-Pech[3,6]*

[1]Agronomic Institute of Paraná, Ponta Grossa-PR, Brazil
[2]SCOPELA, Broissieux, F-73340 Bellecombe en Bauges, France
[3]INRA, UR 0767 Ecodéveloppement, Avignon, France
[4]INRA, UMR1248 Agrosystèmes et développement territorial, F-31326 Castanet-Tolosan, France
[5]Université de Lyon, CNRS UMR 5558, Villeurbanne Cedex, France
[6]Facultad de Medicina Veterinaria y Zootecnia. UADY, Mérida, Yucatán, México

## Abstract

Biodiverse pasture ecosystems, such as rangelands, are now highly valued for their ecological, landscape and agronomic properties. The vegetation of these plant communities usually consists of a variety of

---

\* Corresponding author's email: c.agreil@scopela.fr.

herbaceous and shrubby species. Grazing and browsing by domestic herbivores has been proposed as an economic and efficient way to restore biodiversity and forage resource quality in herbaceous-shrub mosaics by, for example, avoiding shrub encroachment. Further, novel concepts in grazing science draw attention to the importance of maintaining functional heterogeneity in order to sustain food intake through behavioral adjustment. However, the challenge is now to define the heterogeneity, allowing integration of productive goals and the control of shrub dynamics in order to guide practices for rangeland management. Therefore, in this chapter, we aim to analyze the recent conceptual advances in the role of diverse vegetation on intake dynamics. We explore the foraging responses of ruminants faced with a diversity of herbaceous feed items and their effects on shrub (e.g. *Cytisus scoparius* Linck) consumption. We aim also to take into account the role of these feeding choices on intake dynamics at different time scales. Our approach is mainly based on experiments carried out in southern France on ewes grazing relatively small fenced paddocks for short periods of time in shrubby rangelands. Flock activities were always recorded through scan sampling method, and the ewes' diet selection was encoded as bite categories. For instance, the effects of three different herbaceous covers with different forage availability and quality on shrub consumption were compared, mainly by showing how the availability of "herbaceous bites" and selection affect the way that ewes integrate shrubs into their diet. We also identify the behavioral adjustment possibilities between the size and quality of herbaceous *vs.* shrubby bites into meals. The results provide insights into ways to manipulate diet selection in order to stimulate the use of a dominant shrub by ewes. We highlight how grazing management of herbaceous physiognomy can lead to efficient shrub encroachment control. We discuss the usefulness of bite categories as functional feed indicators. Finally, operational implications are discussed in order to implement original approaches that take into account the functional interactions between ecological and technical processes. Adaptive management and its operational implementation are suggested as a promising approach.

# 1. Introduction

There is an increasing interest by both consumers and livestock producers in the development of more sustainable grazing systems, with less dependence on external finite resources (Villalba et al., 2009). With this production perspective, rangelands have been identified as useful vegetation, as they provide diversified resources that increase the resilience of livestock farming systems (Provenza et al., 2003). Thus, biodiverse pasture ecosystems such as rangelands have the potential to meet these challenges. The rangelands that result from grazing and browsing are an important resource world-wide (Gordon et al., 2004). These natural areas are also attracting attention because of their range of conservation designations (Magda et al., 2009; De Bello et al., 2010) and their value for tourism, which is greatly due to their diversity (Agreil et al., 2006).

In pastures with high plant diversity the spatial heterogeneity is often high and the various ecological processes associated with the presence of grazing animals are more complex (Pihlgren & Lennartsson, 2008). Such complexity occurs because the animals both create, and respond to, temporal and spatial variations in the state of the vegetation (Parsons & Dumont, 2003). An important factor frequently associated with the heterogeneity is the increased biodiversity (Adler et al., 2001). Indeed, where different plant life forms are present, particular foraging behaviour patterns become possible. However, from a practical or management perspective, livestock farmers are not very enthusiastic about restoring or conserving "heterogeneity". Since biodiverse communities vary in space and time, these ecosystems become less predictable and hence management difficulties increase. Fortunately, despite the easier control of uniform vegetation or artificial grasslands, there is a recognition that heterogeneity plays an important role in increasing the resilience, adaptability, and productivity of diverse rangeland ecosystems, and thus in increasing the options for plants, herbivores, and people (Provenza et al., 2003).

Although research has increased the interest in using heterogeneous rangelands, the effects of increased diversity on livestock productivity have not been well explored. For instance, previous works have shown that a greater number of choices, both in relation to species richness and food structure, can motivate feeding (Meuret & Bruchou, 1994; Baumont et al., 2000; Agreil et al., 2005), allowing animals to meet their nutrient requirements (Villalba et al., 2009). According to Agreil et al. (2005), in a study on diverse vegetation, within a single day, adjustments in animal feeding behavior occur in order to stabilize the daily average digestibility and bite mass. The herbivores may also change their diet composition simply to avoid wide variations in nutrient content throughout the year (Barroso et al., 1995). These results suggest that animal feeding behavior can vary when there is variability in the diversity of edible items. Therefore, as the size and quality of the resource is highly diverse in more complex rangelands (e.g. a diversity of grasses and shrubs), and the animal response is variable over time (Agreil et al., 2005), alternative management options are constantly required to cope with the changes in these communities.

Within a given plant community, co-occurring plant species may adopt different strategies in order to respond to disturbance. A vast amount of literature has described such strategies (e.g. Westoby, 1999; Papachristou et al., 2005; Díaz et al., 2006) and their effects on ecosystem processes (e.g. Suding et al., 2008), such as biodiversity maintenance and productivity. For instance, avoidance and resistance responses to disturbance, such as changes in canopy size and architecture, are related to ecosystem processes like accumulation of standing biomass (De Bello et al., 2010). Interactions between species also play an important role in influencing the structure of plant communities. In general terms, if the net outcome of these interactions is negative we call it "competition", whereas if the final balance is positive we speak about "facilitation" (Brooker et al., 2008) or "complementarity" (Gross et al., 2007). These biotic mechanisms coexist in semi-natural plant communities and interact with the environment (Malkinson & Kadmon, 2007), producing a wide range of variability, which in turn affects feeding behavior (Adler et al., 2001; Iason & Villalba, 2006). Therefore, grazing management methods cannot be based on a statistical paradigm of equilibrium, ignoring scaling effects in time and space (Laca, 2009). Spatial and temporal heterogeneity of resources needs to be incorporated in order to establish management decisions and methods to control livestock diets and impacts.

Our aim in this chapter is to discuss the current knowledge regarding the complex interactions between vegetation dynamic in biodiverse and heterogeneous communities and the feeding behavior dynamic of herbivores, particularly ruminants. Further, we discuss tools for conceiving adaptive and adjustable management in relationship with a functional view of flock-vegetation-management interactions.

## 2. Ecological Processes Affecting Heterogeneous Systems

By improved understanding of key processes controlling vegetation dynamics at a paddock scale and over time, rangeland managers may be able to achieve better ecological and economic returns. Therefore, in the next part of this section, we present ecological processes which strongly influence the maintenance of diversity in rangelands such as growth (individual scale), demography (population scale) and competition (community scale).

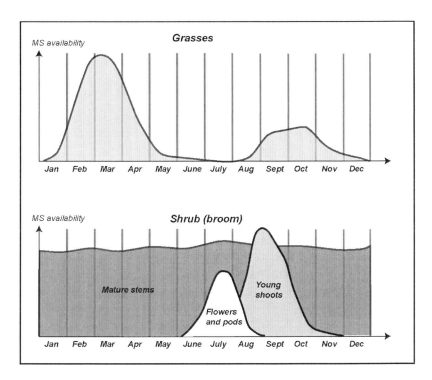

Figure 1. The pattern of available resources in grasses (top) and in shrubs (*Cytisus scoparius* namely broom, bottom) throughout the year (from Agreil et al., 2010).

## 2.1. Growth

A substantial proportion of the world's rangelands are populated by a great variety of grasses and shrubs, which result in various vegetation types (e.g. maquis, garrigues, etc., see Papanastasis et al., 2008 for definitions). Furthermore, the role of these communities in areas with a long dry season and harsh environmental conditions (e.g. Mediterranean shrublands, Sardans et al., 2008) has been widely recognized (Rogosic et al., 2008), because they can fill the periods of feed shortage in the year (Papachristou et al., 2005).

The availability pattern for trees and shrubs is more uniform throughout the year (see Figure 1). Shrubs with green twigs (i.e. young shoots, center Figure 1), whose availability pattern is very stable, become a key resource for grazing. Their resources are accessible even during the non-vegetative growth periods. On the other hand, the pattern of available resources from an herbaceous cover varies over time and depends on its specific composition. In temperate pastures, all herbaceous species, especially grasses, generally include two availability peaks, one in spring and another in autumn (top, Figure 1). Although these growth periods can be extended by a mixture of early and late species, there is a marked lack of growth during summer and winter.

Thus, shrubs and woody species can provide green forage for grazing animals at specific critical periods of the year when herbage is limited (e.g. between June and August, Figure 1). In addition, some shrub species (such as leguminous species of the genera *Cytisus* and *Genista*) can play a significant role in providing protein-rich fodder (Tolera et al., 1997) when shrub twigs are highly palatable to ruminants (Holst et al., 2004). Consequently, in each fenced pasture, by combining the various utilization periods throughout the year (bottom, Figure 1), resource availability can remain compatible with the annual animal feed requirements and farmers' grazing objectives.

Another example of the importance of understanding the growth dynamics of the plant community for planning feed resources, in this case for herbaceous species, is given in Figure 2, which shows the growth patterns of communities composed of two different species with two distinct growth strategies (nutrient acquisition versus conservation). *Brachypodium pinnatuns*, a species with a conservation strategy, is characterized by late growth with a long leaf lifespan and a late flowering period (Al Haj Khaled et al., 2005). Grazing or cutting of communities dominated by this species can be delayed until after the main growth periods. In this case it is possible to use biomass of good nutritive value during periods of low growth (*e.g.* early summer, late autumn). Communities which can maintain a good nutritive value over the spring growing period cope better with different management than those dominated by early species, and give more flexibility for management of the forage system (Duru et al., 2008).

In these former communities, later cutting would have less effect on the ratio of harvested biomass/produced biomass, unlike early species whose growth is rapid and whose biomass needs to be used during a short period in spring. From the point of view of management flexibility, it is possible to take advantage of the ability of communities to tolerate different cutting/grazing regimes to create feed resources which differ in both quality and quantity, according the utilization period. Therefore, regardless of vegetation types and the species' growth strategies, it is important to understand the way that plant populations and communities are structured in order to determine how forage and animal production is affected.

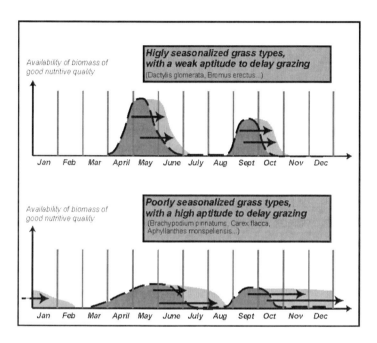

Figure 2. Annual growth distribution and ability to maintain nutritive value after the main growing periods, for two different vegetation types.

## 2.2. Demography

The diversity and heterogeneity of shrubby rangelands are often threatened by shrub dominance and to a lesser extent by herbaceous species, such as *Brachypodium sp.*. Understanding the processes that determine the demography of plant populations will enable us to control the species abundance through grazing.

Population ecology has already made significant progress in explaining the behavior of populations and their demographic dynamics. Mathematical tools can be used to develop models to simulate the demographic dynamics of populations over a relatively long period of time on the basis of knowledge of various demographic processes (reproduction, survival, etc.) and the quantitative estimation of demographic fluxes that contribute to population growth rates (fecundity, survival rates, seed survival, etc.) (Caswell, 2001;

Florian et al., 2008). These analyses provide insights on "the demographic strategy" of the species, i.e. an understanding of the demographic traits (such as the production of large numbers of seeds) or combination of traits (such as high fecundity and long-lived seed-bank) on which the species develops its ability to ensure persistence, and sometimes dominance. Numerous studies on different biological models have described, sometimes very precisely, the diversity of strategies, even for dominant species for which abundance cannot usually be related to a specific trait.

These modeling approaches have also been used to explore the effects of various environmental and biotic factors on population demography by identifying the effects of variations in demographic fluxes on the growth rate (sensitivity analysis) and on changes in population structure, e.g. changes in the relative distribution of plants within the various developmental stages from seedling to juvenile or adult, etc. Analyses were made of the impact of herbivores on a large number of plant species, including woody species, mostly to understand the role of phytophagous insects on plant population dynamics, and more specifically to identify insects that might be used for biological control. In most cases, there were not many predictions of the impact of these herbivores on plant population dynamics, mainly because of the lack of knowledge of the herbivores' feeding behavior. The total consumption is usually estimated by indirectly measuring a quantity of biomass consumed, and by incorporating it in the models as a constant over time (Leimu & Lehtilä, 2006; Kelly & Dyer, 2002; Doak, 1992). This approach ignores variations in the herbivores' feed choices and levels of consumption as a function of time and resource availability. Using the shrub Scotch Broom (*C. scoparius*) as a model species, adaptations of the "classical" approach for plant population representations have been developed in order to accommodate the link between the features of the feeding behavior and the plant demographic processes (Magda et al., 2009). The "classical" representations in plant demography modeling usually describe multi-category structuring with various stages of development, i.e. seedlings, seed, juveniles, adults. Explicit information about the nature of organs present in each stage of development is needed to explore the multitude of demographic parameters modified in relation to the types of organs consumed by grazing. On the basis of this representation, we can identify the plant organs whose consumption through browsing is expected to have major consequences on the population growth process, which should be identified as targeted life stages (Magda et al., 2009). Another level of interlinkage is to introduce the long-term impact of browsing on the demographic behavior of the population. Repeated browsing can strongly affect the demographic structure of the population through reallocation of resources to the vegetative growth process (Figure 3). Juvenilization of an adult plant can be observed, creating a new category of plant whose demographic behavior is radically different from that of the reproductive adults. To accommodate the long-term impact of browsing, it is necessary to add this browsing-induced category to the population structure within the population (Figure 4). Finally the plants which are inaccessible to animals, being too tall to be browsed or located within dense clumps, must be also explicitly distinguished in population representations, as their demographic behavior will obviously differ from that of browsed plants. This implies the need for spatially-explicit modeling to take account of the demographic categories defined by the selective feeding behavior in response to the spatial patterns of plant distribution. These different linkages in the plant demography modeling allowed the spatio-temporal dynamics of a grazed shrub population to be simulated, taking account of features of animal feeding behavior and the interactions between plant and animal processes (Figure 5).

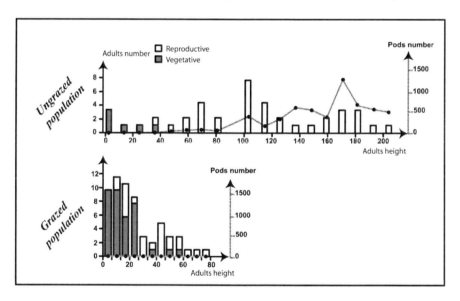

Figure 3. Comparison of the distribution of Scotch broom (*Cytisus scoparius*) adult plants according to their height and reproductive rate for two populations (ungrazed and grazed) showing that grazing reduces the height of adult plants and maintains them in a vegetative state even though they had reached the reproductive age.

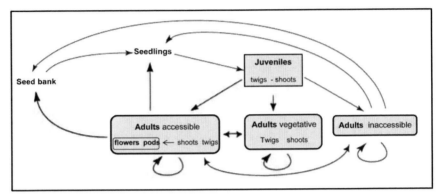

Figure 4. Life cycle of Scotch broom (*Cytisus scoparius*) represented basically by a life-stage structured model with the four main stages: seeds in the seed bank, seedlings, juveniles, and reproductive adults. It has been adapted to link population demographic process with animal feeding behavior processes (i) by identifying organs of each plant stage as feed items (ii) by adding new demographic categories induced by grazing, such as "juvenilized" and "inaccessible" based on the response of juveniles and adults to repeated grazing or browsing (Agreil et al., 2010).

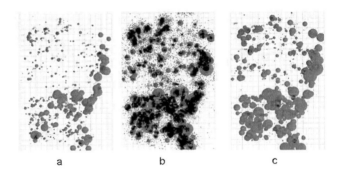

Figure 5. Results of simulation of spatio-temporal dynamics of Scotch broom (*Cytisus scoparius*) from an initial population (a), under no grazing (b) and with grazing management targeting pod consumption (c). Each grey point corresponds to an adult or juvenile plant whose size is proportional to a plant class volume. The black points represent seedlings.

## 2.3. Competition

Brooker et al. (2008) have shown that interactions between herbaceous and vascular plants have been the focus of most plant interaction studies. For instance, Pihlgren & Lennartsson (2008) showed that shrubs may affect grassland plants adversely by competition, positively by serving as grazing refuges (we can speak about "facilitation") or neutrally. Shrubs may also benefit herbaceous plants by, for example, transferring nutrients from deeper soil layers to the surface through leaf litter (Moro et al., 1997). Many aspects (stage of establishment of the shrub, season, water availability, etc.) can influence whether the net effect will be positive or negative in the interaction between shrubs and herbaceous vegetation. Therefore a better understanding of these processes has a direct relevance for the development of tools for livestock activities, and for improving our understanding of the response of plant species and communities to the drivers of environmental change (Brooker et al., 2008). Here we focus particularly on the interactions between herbaceous and shrubby species, since shrubs are viewed as elements increasing heterogeneity of grasslands, and hence plant diversity (Pihlgren & Lennartsson, 2008).

The biotic processes, competition and facilitation, are mediated through changes in the abiotic environment or through other organisms (both plant and animal, Brooker et al., 2008). They may also change over time (Köchy & Wilson, 2000). For example, leguminous shrub seedlings might benefit nearby herbaceous plants by increasing the nitrogen supply due to the presence of seedlings with *Rhizobia* nodules in their roots (Espigares et al., 2004). Conversely, when these shrubs are mature, the conditions for nearby herbs and grasses may be less favorable because of a deeper litter layer (Jensen & Gutekunst, 2003) caused by leaf deposition and reduced grazing intensity and light availability (Pihlgren & Lennartsson, 2008). Jensen & Gutekunst (2003) have reviewed the role of litter effects for species composition. They argued that the litter layer affects the germination of seeds by reducing the amplitude of temperature and light (quantity and quality) (important for breaking dormancy), having chemical effects due to dissolved litter substances (allelopathy) and mechanical effects due to litter mass and density that act as a barrier for germination and establishment. However, they concluded that the reduction of seedling recruitment by a litter layer is also species-specific and that aspects like seed mass or seedling morphology must be included in these analyses.

Although the shrubs benefit the herbs during the establishment of the shrubs, the shrubs themselves suffer from the association during this phase (Espigares et al., 2004). Herbaceous plants compete with shrub seedlings by limiting the survival and growth of the shrub (Espigares et al., 2004) and by reducing water availability (Clary et al., 2004; Pataki et al., 2008). These adverse effects are exacerbated when the beginning of shrub establishment coincides with the maximum growth of the herbaceous vegetation, as in spring. Thus, early emergence may avoid competition with herbs and increasing water inputs could also minimize the harmful effect of competition (Espigares et al., 2004). However, once water becomes available, opportunistic grass species with a water maximization strategy could consume soil moisture quickly after drought and hence reduce the availability of water to slower-responding plants growing within their sphere of competition (Clary et al., 2004).

As a result, these findings suggest that during early stage of shrub expansion, competition for soil resources is the most important factor and the allocation patterns seems to contribute to competitive ability. As seedlings increase in mass and height, they should become more competitive and the competition for light becomes the dominant process (Köchy & Wilson, 2000). By then the shrubs may become dominant, resulting in a substantial loss of biodiversity and forage resources in pastoral areas (Bartolomé et al., 2005; Dalle et al., 2006; Kesting et al., 2009).

Grazing by domestic herbivores has been proposed as a major way to control shrub encroachment (Valderrábano & Torrano, 2000; Bellingham & Coomes, 2003; Pontes et al., 2012). However, intense clearing of shrubs in biodiverse pastures cannot be a general recommendation, since several benefits result from the heterogeneity and various interactions that take place in grass-shrub mosaics. However, achieving good grazing management of these communities presupposes knowledge of how ruminants interact with the vegetation, which appears to be related to the ability of herbivores to move and access key functional vegetation resources.

## 3. Feeding Behavior in Heterogeneous Communities

Diet selection by ruminant herbivores is influenced by many factors. Among them are learning and the different time scales considered in the dynamics of ingestion behavior. Here we review behavioral mechanisms and challenges ruminants face when foraging heterogeneous plant communities. We consider the most relevant time scales for describing feeding adjustments according the vegetation changes. It is important to understand how herbivores make their choices when selecting foods from very heterogeneous vegetation in order to produce management guidelines in grazing ecosystems.

## 3.1. Learning

Understanding plant-herbivore interactions increases potential implementation of approaches for sustainable management of ruminants in heterogeneous vegetation. Learning by animals is an important mechanism resulting from these interactions, because it plays a key role in an animal's propensity to eat foods that differ in amounts of nutrients and toxins (Provenza et al., 2003). This individuality has significant implications for managing a heterogeneous environment and needs to be considered in grazing and browsing management.

According to Hessle (2009), animal behavior can be environmentally manipulated through learning. For instance, spatial learning of new foraging areas may be enhanced by using experienced animals to act as social models. When in a new environment, untrained grazers may adopt a modest foraging behavior, resulting in suboptimal pasture management and live-weight gains. When experienced animals accompany them, the untrained grazers may be more efficient in finding preferred food on heterogeneous pastures, increasing their focus on a smaller number of plant species with a higher bite rate (Ganskopp & Cruz, 1999; Ksiksi & Laca, 2000). However Hessle (2009) failed to improve foraging ability in cattle by using experienced cattle, in terms of increased live weight gains. This author suggested that there may be some benefit from social learning in applied pasture management, by making the herd calmer and easier to handle. Meuret et al. (2006) showed that livestock farmers can develop original management for diary heifers, in order for them to learn to walk on steep slopes.

Ruminant herbivores also learn about the nutritional properties of food by associating their post-ingestive consequences with their sensory experiences (Provenza, 1995). Feedback during digestion can be either positive (improved nutrient or energy status of the animal) or negative (illness due to over-ingestion of toxins or nutrients), rendering the plant flavor more desirable or aversive (Papachristou et al., 2005). Furthermore, Kronberg & Walker (2007) showed that the ability of sympatric ruminant species to learn to decrease or increase intake of a particular plant is likely correlated with how their morphophysiology interacts with plant chemicals. For instance, sheep can tolerate and detoxify more of some alkaloids than cattle (Launchbaugh et al., 2001). These facts are particularly important in rangelands, in which most plants contain some secondary metabolites.

In heterogeneous vegetation, animals encounter numerous plant species that differ in their structure and their nutrient or toxin content. In these environments, animals can learn to mix and select a diet which is richer in nutrients and lower in toxins than the average available resource (Illius & Gordon, 1991). Further, herbivores often consume plant species and parts that do not maximize rates of intake (see review by Papachristou et al., 2005). Although they may not select the best quality diet available, they may avoid wide variations in nutrient content or toxin in their diets throughout the year (Kababya et al., 1998). Thus, previous animals' experiences in combining different foods may enhance diet breadth and may promote greater use of all plants in heterogeneous ecosystems over time. Grazing management practices need, therefore, to encourage animals to use different plants of a rangeland in order to mix a variety of nutrients and toxins. In such situations of high diversity, the complementarity between resources will likely help herbivores to avoid toxicity, to have adequate availability of nutrients, and also will help them to reach physical satiety at different time scales (Agreil et al., 2005).

### 3.2. Paddock Scale

Particular feeding behavioral patterns have been observed at the paddock scale (*i.e.* the sequence of days spent within a single fenced paddock during rotational management). Experiments were conducted with flocks of ewes, grazing for a sequence of 7-5 days in small and fenced pastures in southern France. This led to very rapid changes in the available resources and nutritional quality. Agreil et al. (2005) identified the behavioral adjustments of ewes for coping with heterogeneous and variable availability. The authors showed that when ewes are faced with a diverse supply, they increase their range of bite sizes as the size and structure of previously grazed plants decreases, thus stabilizing daily average bite mass and intake rate (Figure 6). The classical behavioral adjustment (the compensation of bite mass decrease by an increase of bite frequency, Allden & Whittaker, 1970) is then replaced in diversified vegetation by a stabilization pattern (the compensation of plant structural changes by an enlargement of bite mass range, Agreil et al., 2006).

On rangelands, besides choices among species, ruminants also have to choose their feed among plant organs within the species, e.g. twigs, leaves and fruits (Agreil & Meuret, 2004). Thus, in order to maintain the stability of daily intake, a variable use of plant organs within species can be observed. For instance, with a shrubby species like broom, the choice criteria may be between small bites in more palatable organs (flowers and pods) and large bites on long twigs (Agreil et al., 2010).

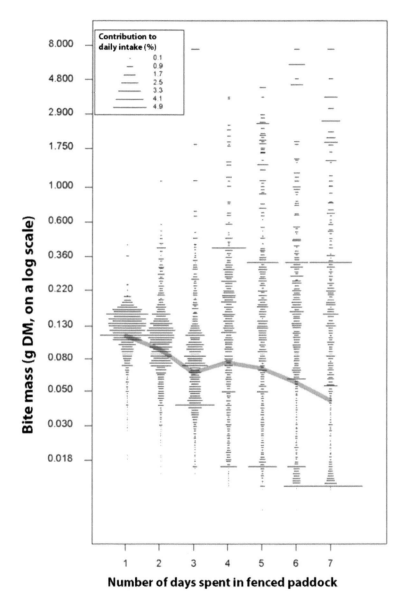

Figure 6. Day-to-day variation of the daily average bite mass (black line) and the distribution of instantaneous bite masses during the paddocking sequence (7 days). The length of the horizontal black dashes is proportional to the average contribution of each bite mass category to the total daily dry matter intake (%). The gray line indicates the daily average bite mass variation during the time spent in the fenced paddock. Figure from Agreil et al. (2006).

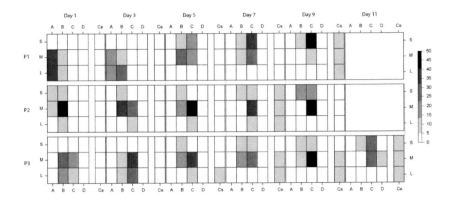

Figure 7. Percentage of ewes in each vegetation bite category for each sampling day (columns), and each paddock (lines, P1, P2 and P3, respectively). P1, 100% previously grazed paddock in summer; P2, 50% previously grazed paddock in summer; P3, paddock not previously grazed. A, 100% of green leaves; B, 100% of green tissue (leafs and stem); C, for bites comprising less than 70% dry tissue; and D, more than 70% dry tissue. Cs represents *Cytisus scoparius*. L, M and S represent the bite mass: L, large bites; M, medium-sized bites; and S, small bites. From Pontes et al. (2010).

This temporal pattern of diversified bite mass consumption has been also reported in a study on broom (*C. scoparius*) shrubland (Pontes et al., 2010). In this study, small paddocks consisting of multi-stratified vegetation dominated by herbaceous species encroached by broom (a perennial long-lived leguminous shrub), were grazed by ewes for 10 days in the autumn. At the beginning of paddock utilization, ewes maintained lower ranges of bite masses, i.e. they were highly selective because the food sources were diverse and abundant (Figure 7). As soon as the availability of larger bites of higher nutritive value in herbaceous vegetation declined, ewes gradually started to select large bites of broom together with smaller bites of herbaceous species. Thus ewes may take smaller and more nutritious bites of herbaceous plants and larger bites of broom in order to maximize daily energy intake.

### 3.3. Daily Intake

The bite coding grid method, i.e. a direct observation and recording of bites, has often been used to characterize the feeding behavior of ruminants (e.g. Agreil & Meuret, 2004; Pontes et al., 2010; González-Pech & Agreil, 2012). In Figure 8, daily intakes measured in four experiments were plotted against the daily mean organic matter digestibility of the diet, and compared to data from the literature (reference model, Figure 8). Therefore, thanks to the coding grid method, we can observe (see Figure 8) a considerably higher daily intake of dry matter and nutrient content of the diet than those predicted in the models published by Morley (1981) and Van Soest (1994). These results suggest a positive effect of the diversity on forage available at pasture.

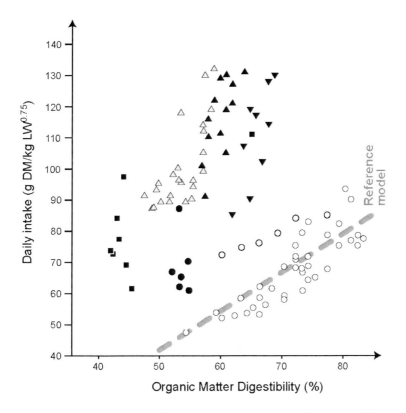

Figure 8. From Agreil & Meuret, 2004. Relationship between daily intake of DM and organic matter digestibility of the daily diet. Each solid symbol represent a single observation of daily intake by an individual: lactating goats herded during the summer in oak coppice (black triangles on their base; Meuret et al., 1994; Meuret, 1997; Baumont et al., 2000); lactating goats fed fresh leafy oak branches frequently renewed in digestibility crates (open triangles; Meuret, 1988; Meuret & Giger-Reverdin, 1990); dry ewes grazed in early spring on natural swards encroached by edible shrubs (black circles; Agreil & Meuret, 2004) and dry ewes grazed in summer on natural swards encroached by edible shrubs (black squares; Agreil & Meuret, 2004). The dotted line represents the reference models published by Morley (1981) and Van Soest (1994).

## 3.4. Meal

A second important temporal scale is the meal, defined as the time ruminants devote to ingestive behavior, and separated by two resting or ruminating periods. Meuret et al. (1994), for instance, detected a positive effect of diversity on intake rate during meals. Agreil et al. (2008) reported cyclic patterns within intake rate kinetics during feeding bouts (Figure 9). The feeding bout experimental variograms were satisfactorily modeled by a sum of three basic models: a long-range exponential model, a damped cosine model, and a short-range exponential model. Intake rate time series can thus be interpreted as the sum of three components: 1°/ an exponentially decreasing dependence, which can be considered negligible for lags longer than 20 min; 2°/ correlation coefficients that are successively positive and negative, generating an oscillating pattern, with pseudo-periods of 19.6 and 14.7 min for experiments 1 and 2 respectively; 3°/ an exponentially decreasing dependence, assumed to be negligible for lags longer than 2 min. When foraging within their flocks, individual animals probably do not have complete information about available plant parts at any given moment (Fortin, 2002), and their metabolic consequences (Villalba & Provenza 2005). Oscillations, which are frequent in biological systems (Slater, 1999), are sustained and not damped in our case. Such intake rate oscillations have already been identified in a grazing mollusc foraging on a variety of resources (Chelazzi et al., 1998). The temporal scale of the oscillating pattern (pseudo-period about 20 min) could be analyzed with regard to the kinetics of elimination of the main toxins, and in particular their half-life profiles.

## 3.5. Bite/Instantaneous Feeding Behavior

The direct observation method presented by Agreil & Meuret (2004) was used in situ to record the ingestive behavior of animals on diverse vegetation and at different time scales, from the bite level (a few seconds) to the paddocking sequence (several days). From these observations, they defined a bite-coding grid in which bite categories are common to several species, on the basis of the structure of the removed material. The detailed description of bite selection obtained from this bite coding grid method enabled us to investigate the link between bite mass and frequency. We present here the results obtained in an experiment with Brome grass (*B. erectus*, Figure 10) which was described by Agreil & Meuret (2004). In this work, 15 bite categories were applied to Brome grass for 15 days in a paddock. The bite mass observed was between 0.007 and 0.142 g dry mass (DM), and bite frequency varied considerably from 10 to 100 bites.min.$^{-1}$. The general trend reflects a decrease in bite frequency with increasing bite mass (Figure 10), a result that has already been well documented in the literature. This effect was observed in particular for long sequences of uninterrupted consumption of Brome (more than 3 min.). Frequencies above 60 bites.min.$^{-1}$ were mainly associated with bite masses below 0.025 g DM, and resulted in intake rates which rarely exceeded 2 g DM/min. On the other hand, intake rates above 3 g DM/min. were mainly observed for bite frequencies below 60 bites.min.$^{-1}$ and bite masses above 0.07 g DM. Therefore, an increase in bite mass increased intake rate, whereas increased bite frequency appears to be less efficient. This effect is generally attributed to the time needed to harvest and masticate plant organs (Spalinger & Hobbs, 1992).

Another example of the relationship between bite frequency, bite mass and intake rate is shown in Figure 11. This study was made in the Crau steppe, an environment with a unique herbaceous layer and a lot of *Brachypodium retusum*. The figure shows that higher values for intake rate (around 11 – 13 g DM.min.$^{-1}$) are possible when ruminants combine bite masses of 0.150 g of DM with bite frequencies between 40 – 70 bites.min.$^{-1}$. These values are higher than those reported in the literature, which are around 4 – 6 g DM.min.$^{-1}$ and are thought to be the maximum possible by sheep when grazing temperate herbaceous vegetation (e.g. Iason et al., 2000). A high intake rate is generally associated with large bites. For instance, Meuret (1997) report that intake rates greater than 13 g DM.min.$^{-1}$ are possible when grazing shrubs, which provide bite masses of around 1g DM, but with low bite frequencies, around 15 – 17 bites.min.$^{-1}$. Therefore, despite the absence of shrubs in the Crau steppe, high intake rates were observed. Further, this study showed that when *B. retusum*, considered to be of low nutritional value and little appreciated by the sheep, contributes the majority of dry matter intake, intake rate is significantly greater (6.8 g DM.min.$^{-1}$) than when it comes from forbs (4.4 g DM.min.$^{-1}$) or other grasses (5.8 g DM.min.$^{-1}$). Hence *Brachypodium* should be regarded as an important feed resource, since it facilitates instantaneous intake. In conclusion, this kind of vegetation, with only herbaceous species, can be considered as a food source similar to woody or shrubby rangelands, thanks to a species that allows large bites, increasing the efficiency of ingestion.

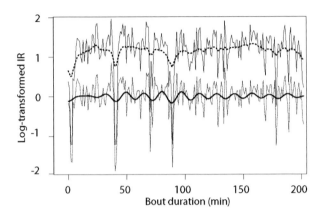

Figure 9. Filter kriging during one bout, as an example. The log-transformed intake rate time series (thin solid line) is represented with the filtered trend (thick dotted line), the oscillating component (thick solid line) and the short-range (thin dotted line). Figure from Agreil et al. (2008).

## 3.6. Conclusion Regarding Animal Processes

Many of the results presented in this section may contribute to the current debates in domestic ruminant feeding science and in nutritional ecology, concerning the choice of variables and "objective functions" to be included in models. In the optimization models, maximization is the function used by far most (Stephens & Krebs, 1986), together with the variables that describe ingestion efficiency (rate of intake, nutritive value, etc.). The debate on the validity of these models and how to depict ruminant behavior remains open because of the continued inflow of new field data (see review by Sih & Christensen, 2001).

A schematic representation of the consequences of two different feeding strategies is given in Figure 12. The volumes shown in grey represent the range of intake rates and digestibility of the dry matter ingested each day. In this figure, the darker the gray, the greater the contribution. In the case of the maximization function (Figure 12, left), a real maximization is never observed (Illius et al., 1999), and the daily volumes are very widely spread, with a darker gray in the upper right-hand side. When ruminants are kept in a fenced pasture, the intake maximization hypothesis leads to a gradual depletion of high quality food and to a shift in the range used towards lower intake rates and digestibility (see the arrow in Figure 12). Conversely, and according the studies described above, a different hypothesis can be proposed (see Figure 12, on the right) within optimization models, i.e. a stabilization function hypothesis. This hypothesis is also represented in Figure 12, which show that the center of gravity of the daily ranges explored remains in the center of the space described by the intake and digestibility range used.

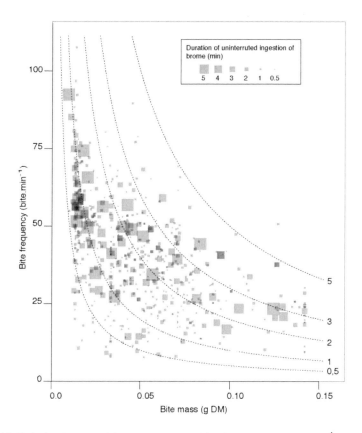

Figure 10. Relation between bite mass (g DM), bite frequency (bites.min.$^{-1}$) and intake rate (g DM.min.$^{-1}$) for a ewe eating brome grass (*Bromus erectus* L.) (Agreil & Meuret, 2004). Each dot represents a sequence of uninterrupted ingestion of brome grass. The position of the dot in relation to the horizontal axis indicates the average bite mass (g DM) of the sequence. The position of the dot in relation to the vertical axis indicates the average bite frequency (bites.min.$^{-1}$). The dot size is proportionate to the duration of the sequence (min.). Their position in relation to the isoclines, which connect dots whose coordinates reflect ingestion that occurs at the same rate (g DM.min$^{-1}$, to the right), provides information on the average intake rate during each of the sequences. Figure from Agreil & Meuret (2004).

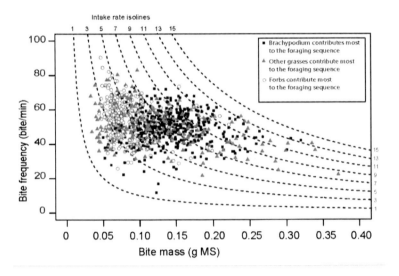

Figure 11. Relation between bite mass (g DM), bite frequency (bites/min.) and intake rate (g DM/min.) for ewes grazing in the Crau steppe (adapted from González-Pech & Agreil, 2012). Each dot represents a sequence of uninterrupted ingestion when forbs (white circles), *Brachypodium* (black squares) and other grasses (gray triangles) contributed the most to the intake. Dots' positions in relation to the intake rate isolines provide information on the average intake rate (g/min.) during each one of the sequences.

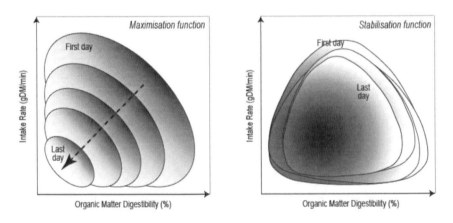

Figure 12. Schematic representation of the consequences of two different functions used in modeling feeding strategies: maximization function (left) and stabilization function (right). The gray volumes represent the range of intake rates and digestibility of the organic matter ingested each day, the darker the gray, the greater the contribution (Agreil et al., 2006).

This feeding behavior can be interpreted as a response of animals in order to cope with big variations in food supply, particularly in heterogeneous vegetation. Such stabilizing behavioral adjustments may be of particular interest for grazing conditions in which the individual animal can never obtain full information on the range of available feed or the pace of its depletion, because the animals in the group all eat together, at the same time, and in the same enclosure.

## 4. Management in a Heterogeneous and Dynamic System

Understanding the farmer-animal-plant interactions is crucial for managing plant and herbivore populations in heterogeneous vegetation. Herbivores interact with a food resource that is heterogeneous on different spatial scales. Grazing influences the vegetation structure and species composition on the landscape level, and chemical and morphological properties between and within individual plants, promoting changes in the subsequent regrowth. Further, herbivores need to develop behavioral strategies to overcome these plant responses. Positive and negative feedback can be observed. For instance, a reduction in senescent material and maintenance of plants in an early phenological state encourages the continued use of previously grazed patches, i.e. a positive feedback (Adler et al., 2001). On the other hand, physical defenses may be effective deterrents against herbivores (negative feedback). Therefore, farmers need to deal constantly with a very dynamic pastoral system. A simple schematic representation of this complex system is proposed in Figure 13. Complexity emerges because each of these three components is itself governed by its own biological or biotechnical processes, and because the interactions with the two others affect these intrinsic dynamics.

Some examples of such an interaction between these three components of a pastoral system (Figure 13) are given for our experiments on rangelands encroached by Scotch broom shrubs and grazed by ewes. In order to better explain the interactions between these components, the description of the plant community included the functions of the various edible plant organs according to their structure, that determines whether the ruminants will take either large or small bites. For instance, in a study by Pontes et al. (2010), in the absence of large ($0.37 \pm 0.105$ g DM) and medium ($0.16 \pm 0.051$ g DM) bites in herbaceous plants with 100% of green tissue, ewes started to browse broom shrubs (see Figure 7). Thus, the abundance, size and quality of these bite categories provides qualitative indicators of the state of herbaceous cover and whether it can trigger initial shrub consumption. In order to characterize the effect of grazing management, some paddocks were not grazed during the main growing period (spring and summer). In this case, the bites offered by herbaceous cover during the following fall were of medium or low palatability (as a result of an increased development of stems and spikes and a lower leaf:stem ratio). In these paddocks, ewes were encouraged to include broom target parts in their diet. They selected large bites of broom at the same time as they selected smaller and more nutritious bites of herbaceous plants, probably in order to maintain daily energy intake. This finding indicates that resource managers can manipulate herbaceous cover to stimulate the consumption of broom by ewes. However, in order to modify the demographic behavior of this shrub, this management strategy must be applied over several successive years (Pontes et al., 2012).

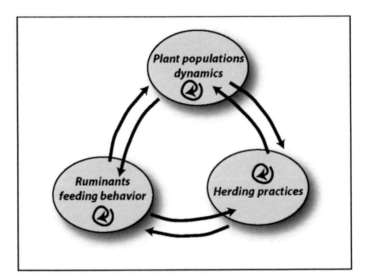

Figure 13. Schematic representation of the three components of a pastoral system and their interactions (adapted from Agreil et al., 2011b). Each of these components simultaneously has an intrinsic dynamics, and affects and is affected by the two others. This confers complex properties to the system, and argues for adaptive management methods in which the effects of herding on grazing behavior and vegetation dynamics are not known in advance.

Magda et al. (2009) showed that juvenile survival is the key demographic parameter for shrub population dynamics. Thus, at the time when juvenile broom twigs (on plants aged >1 year until the reproductive stage, i.e. 3 – 4 years) are easy to target, they are also the plant organs that browsers need to consume to effectively influence shrub population dynamics. In broom, juvenile individuals represent a smaller size and biomass quantity. Further, they can play a significant role in providing protein-rich fodder (Tolera et al., 1997), mainly at early stages, when broom twigs are highly palatable to ruminants (Holst et al., 2004). Therefore, with a previous strategic herbaceous ground cover management, broom juveniles can be easily targeted without excessive grazing over time, which could reduce community diversity. It is important to emphasize that the challenge is to control the growth rate of shrub populations rather than eradicate the dominant species, such as broom, since they have an important role in intake organization by ruminants, and could generate synergies when included within mixed diets.

The complexity of grazing systems (i.e. interactions between plant dynamics, animal foraging behavior and livestock management practices) necessitates formal guidelines for land managers and livestock farmers. Reintegrating biological diversity and ecological system behavior into the production process requires us to increase our awareness of the uncertainties and complexity of the system to be managed. For this reason, adaptive management framework, originally conceived for ecosystem management (Holling, 1978; Johnson, 1999), has been increasingly used in various agricultural systems. For the case of grazing management there are many examples (Janssen et al., 2000; Gross et al., 2006; Launchbaugh, 2006). A case of operational implementation of such grazing adaptive management is given by Agreil et al., 2011a (Figure 14). Five steps are proposed: 1. Agro-ecologic characterization of vegetation, with an explicit recognition of the positive role of diversity. 2. Characterization of livestock farmers' feeding strategy, with an explicit recognition of the different sequences during the year. 3. Agro-environmental objective formalization, seen as a key step in order to target the planning of grazing management. 4. Designing grazing management practices, with an explicit programming of output states to be obtained by grazing at the plot scale and of the management adjustments to be implemented. 5. The operational implementation of grazing management.

Concrete examples of such targeted adaptive grazing management are given in Figure 15 for the maintenance of river bank stability and in Figure 16 for the control of shrub demographic dynamics (*Rosa canina*).

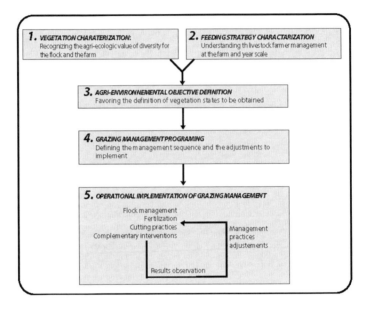

Figure 14. Schematic representation of the 5 steps proposed for implementing the "PaturAjuste" approach (Adapted from Agreil et al., 2011a).

Figure 15. Targeted grazing of riparian areas in Hardware ranch (Utah, USA) maintains bank stability and reduces erosion due to surface flow (Picture by Haskell J.).

Figure 16. Targeted grazing of high altitude meadows (France) by horses grazing keep shrubs (*Rosa Canina*) demography under control (Picture by Agreil C.).

Adopting an adaptive approach for grazing management involves the ability to design, program and implement original management scenarios, which are often far from the usual rough management plans (annual stocking rates, seasonal grazing permits, continuous *vs.* rotational grazing, etc.). As shown in the example above, the choice of season is an important management factor. The fenced pasture utilization period (season, duration) must be matched with the animals' attraction to the targeted plant organs (Agreil et al., 2010). Through his choice of season for pasture grazing, the livestock farmer decides on the vegetation state at the time of entrance, and hence decides to put the flock in a plant community with a given structural and phenotypic state, characterized by certain assemblages of plant organs (see Figure 17). Knowing the time of year when the various target plant organs are produced and palatable would help the livestock farmer decide on the timing of the grazing season. But it is also important to consider the phenology of the other species that could encourage or dissuade animals to eat the target plant organs. The diagnosis of functional feeds (i.e. based on harvest/intake opportunities for animals) available at the beginning of and during the utilization period of a pasture would thus enable the livestock farmer to schedule and adjust the length of stay in a pasture so as to ensure continuous satisfactory intake levels. Through his choice of an end date for a grazing period, he decides the exit vegetation state, which is the criterion for anticipating the subsequent shrub demographic responses. At the end of each utilization period of a fenced pasture, the diagnosis should be focused on the consumption rate of targeted life stages and targeted plant organs, as a prediction of the impact of grazing on the growth of the dominant species.

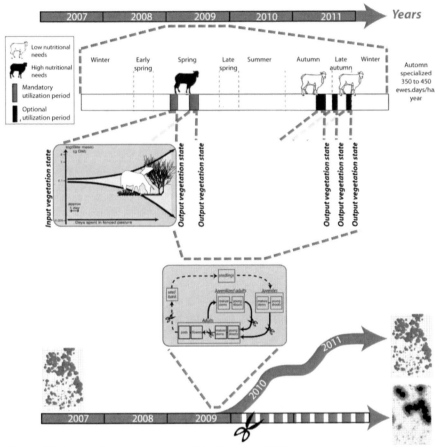

Figure 17. Schematic representation of the functional links between grazing practices and the two main biological processes at the fenced pasture scale. From top to bottom: grazing management results in multiple utilization periods during each year. Each utilization period is characterized by the phenological season, the nutritional needs and a targeted utilization intensity to be applied to the fenced pasture. As a result of this grazing management, ruminants develop their foraging strategies, which in turn determine the grazing impact on each species. At a one-year scale, the utilization sequence has to be summarized as a succession of vegetation states in order to predict the resulting impact on a dominant shrub species. This impact should be considered in terms of its potential effects on the different demographic parameters that constitute the life cycle. Finally, the importance of the successive impacts on the different demographic parameters, and their repetition over the years, will determine the trajectory of the shrub population dynamics in the fenced pasture (Agreil et al., 2011b).

# References

Adler, P. B., Raff, D. A. & Lauenroth, W. K. (2001). The effect of grazing on the spatial heterogeneity of vegetation. *Oecologia, 128*, 465–479.

Agreil, C., Barthel, S., Barret, J., Danneels, P., Greff, N., Guerin, G., Guignier, C., Mailland-Rosset, S., Magda, D., Meignen, R., Mestelan, P. & de Sainte, Marie, C. (2011a). "La gestion pastorale des milieux naturels: mise en oeuvre des MAE-t et gestion adaptative avec la démarche PATUR'AJUSTE", *Fourrages, 208*, 293-303.

Agreil, C., Fritz, H. & Meuret, M. (2005). Maintenance of daily intake through bite mass diversity adjustment in sheep grazing on heterogeneous and variable vegetation. *Applied Animal Behaviour Science, 91*, 35-56.

Agreil, C., Guérin, G., Magda, D. & Mestelan, P. (2011b). Grazing management on dynamic, heterogeneous rangelands: evolution of referential methods at the Regional Park of Massif des Bauges, northern Alps, France. In: T., Kamili, B., Hubert, J.F. Tourrand, (Eds). A paradigm shift in livestock Management: from Resources Sufficiency to Functional Integrity "A workshop of the XXIth Int. Grassland Congress and VIIIth Int. Rangeland Congress, Hohhot, China, 28-29th June 2008. Cardère édition, France. 183-214.

Agreil, C., Magda, D., Guérin, G. & Mestelan, P. (2010). Evolution of referential methods for grazing management on dynamic, heterogeneous rangeland: the case of the Regional Park of Massif des Bauges in the northern Alps. In: H., Hubert, J. F., Tourrand, T. Kamili, A shift in Natural Resources Management Paradigm: from Resources Sufficiency to Functional Integrity? Quae, Paris, France.

Agreil, C. & Meuret, M. (2004). An improved method for quantifying intake rate and ingestive behaviour of ruminants in diverse and variable habitats using direct observation. *Small Ruminant Research, 54*, 99-113.

Agreil, C., Meuret, M. & Fritz, H. (2006). Adjustment of Feeding Choices and Intake by a Ruminant Foraging in Varied and Variable Environments: New Insights from Continuous Bite Monitoring. In: Bels, V. (Ed.), Feeding in Domestic Vertebrates. CABI Publishing, Cambridge, 302-325.

Agreil, C., Monestiez, P. & Villalba, J. (2008). Intake rate oscillations at the meal scale: the dynamics of feeding choices for coping with diversity on rangelands. *In:* Multifunctional grasslands in a changing world. XXI International Grassland Congress. VIII International Rangeland Congress. 26$^{th}$ June – 6$^{th}$ July, 2008. *Huhhot*, China. Vol *1*. 440.

Al Haj Khaled, R., Duru, M., Theau, J. P., Plantureux, S. & Cruz, P. (2005). Variations in leaf traits through seasons and N-availability levels and its consequences for ranking grassland species. *Journal of Vegetation Science, 16,* 391-398.

Allden, W. G. & Whittaker, I. A. McD. (1970). The determinants of herbage intake by grazing sheep: The interrelationship of factors influencing herbage intake and availability. *Australian Journal of agricultural Research, 21,* 755-766.

Barroso, F. G., Alados, C. L. & Boza, J. (1995). Food selection by domestic goats in Mediterranean arid shrublands. *Journal of Arid Environments, 31,* 205-217.

Bartolomé, J., Lopez, Z. G., Broncano, M. J. & Plaixats, J. (2005). Grassland colonization by Erica scoparia (L.) in the Montseny Biosphere Reserve (Spain) after land-use changes. *Agriculture, Ecosystems & Environment, 111,* 253-260.

Baumont, R., Prache, S., Meuret, M. & Morand-Fehr, P. (2000). How forage characteristics influence behaviour and intake in small ruminants: a review. *Livestock Production Science, 64*(1), 15-28.

Bellingham, P. J. & Coomes, D. A. (2003). Grazing and community structure as determinants of invasion success by Scotch broom in a New Zealand montane shrubland. *Diversity and Distributions, 9,* 19-28.

Brooker, R. W., Maestre, F. T., Callaway, R. M., Lortie, C. L., Cavieres, L. A., Kunstler, G., Liancourt, P., Tielbörger, K., Travis, J. M. J., Anthelme, F., Armas, C., Coll, L., Corcket, E., Delzon, S., Forey, E., Kikvidze, Z., Olofsson, J., Pugnaire, F., Quiroz, C. L., Saccone, P., Schiffers, K., Seifan, M., Touzard, B. & Michalet, R. (2008). Facilitation in plant communities: the past, the present, and the future. *Journal of Ecology, 96,* 18–34.

Caswell, H. (2001). *Matrix population models.* Sunderland, MA, USA (2nd ed.), Sinauer Association, Inc. Publishers, 727p.

Chelazzi, G., Parpagnoli, D. & Santini, G. (1998). A satiation model for the temporal organisation of grazing limpets. *Functional Ecology, 12*(2), 203-210.

Clary, J., Savé, R., Biel, C. & Herralde, F. (2004). Water relations in competitive interactions of Mediterranean grasses and shrubs. *Annals of Applied Biology, 144,* 149-155.

Dalle, G., Maass, B. L. & Isselstein, J. (2006). Encroachment of woody plants and its impact on pastoral livestock production in the Borana lowlands, southern Oromia, Ethiopia. *African Journal of Ecology, 44,* 237-246.

De Bello, F., Lavorel, S., Diaz, S. & Harrison, P. A. (2010). Towards an assessment of multiple ecosystem processes and services via functional traits. *Biodiversity Conservation*, *19*, 2873-2893.

Díaz, S., Lavorel, S., McIntyre, S., Falczuk, V., Casanoves, F., Milchunas, D. G., Skarpe, C., Rusch, G., Sternberg, M., Noy-Meir, I., Landsberg, J., Zhang, W., Clark, H. & Campbell, B. D. (2006). Plant trait responses to grazing – a global synthesis. *Global Change Biology*, *12*, 1–29.

Doak, D. F. (1992). Lifetime impacts of herbivory for a perennial plant. *Ecology*, *73*, 2086-2099.

Duru, M., Cruz, P. & Magda, D. (2008). La conduite des couverts prairiaux, source de flexibilité. In: L'élevage en mouvement. Flexibilité et adaptation des exploitations d'herbivores, B. Dedieu, E. Chia, B. Leclerc, C.H. Moulin, Tichit (Eds) Editions Quae, Update Sciences & Technologies, II-3, 57-71.

Espigares, T., López-Pintor, A., Benayas, J. M. R. (2004). Is the interaction between *Retama sphaerocarpa* and its understorey herbaceous vegetation always reciprocally positive? Competition–facilitation shift during *Retama* establishment. *Acta Oecologica*, *26*, 121–128.

Florian, J., Moloney, K. A., Schurr, F. M., Köchy, M. & Schwager, M. (2008). The state of plant population modelling in light of environmental change. *Perspectives in Plant Ecology, Evolution and Systematics*, *9*, 171-189.

Fortin, D. (2002). Optimal searching behaviour: the value of sampling information. *Ecological Modeling*, *153*(3), 279–290.

Ganskopp, D. & Cruz, R. (1999). Selective differences between naive and experienced cattle foraging among eight grasses. *Applied Animal Behaviour Science*, *62*, 293–303.

González-Pech, P., Agreil, C. (2012). Caracterización de la ingestion por observación directa en rebaños ovinos del surest de Francia. *Arch. Zootec.*, *61*, (235), 343-354.

Gordon, I. J., Hester, A. J., Festa-Bianchet, M. (2004). The management of wild large herbivores to meet economic, conservation and environmental objectives. *Journal of Applied Ecology*, *41*, 1021–1031.

Gross, J. E., McAllister, R. R. J., Abel, N., Stafford Smith, D. M. & Maru Y. (2006). Australian rangelands as complex adaptive systems: A conceptualmodel and preliminary results. Environmental *Modelling & Software*, *21*, 1264-1272.

Gross, N., Suding, K. N., Lavorel, S. & Roumet, C. (2007). Complementarity as a mechanism of coexistence between functional groups of grasses. *Journal of Ecology*, *95*, 1296-1305.

Hessle, A. K. (2009). Effects of social learning on foraging behaviour and live weight gain in first-season grazing calves. *Applied Animal Behaviour Science*, *116*, 150–155.

Holling, C. S. (ed.) (1978). Adaptive Environmental Assessment and Management. Chichester: Wiley.

Holst, P. J., Allan, C. J., Campbell, M. H., Gilmour, A. R. (2004). Grazing of pasture weeds by goats and sheep. 2. Scotch broom (*Cytisus scoparius*). *Australian Journal of Experimental Agriculture*, *44*, 553-557.

Iason, G., Sim, D. & Gordon, I. (2000). Do endogeneous seasonal cycles of food intake influence behaviour and intake by grazing sheep? *Functional Ecology*, *14*, 614-622.

Iason, G. R. & Villalba, J. J. (2006). Behavioral Strategies of Mammal Herbivores Against Plant Secondary Metabolites: The Avoidance–Tolerance Continuum. *Journal of Chemical Ecology*, *32*, 1115–1132.

Illius, A. W. & Gordon, I. J. (1991). Prediction of intake and digestion in ruminants by a model of a rumen kinetics integrating animal size and plant characteristics. *Journal of Agricultural Science*, *116*, 145-157.

Illius, A. W., Gordon, I. J., Elston, D. A. & Milne, J. D. (1999). Diet selection in goats: A test of intake-rate maximization. *Ecology*, *80*, 1008–1018.

Janssen, M. A., Walker, B. H., Langridge, J. & Abel, N. (2000). An adaptive agent model for analysing co-evolution of management and policies in a complex rangeland system. *Ecological Modelling*, *131*, 249–268.

Jensen, K. & Gutekunst, K. (2003). Effects of litter on establishment of grassland plant species: the role of seed size and successional status. *Basic and Applied Ecology*, *4*, 579–587.

Johnson, B. L. (1999). The Role of Adaptive Management as an Operational Approach for Resource Management Agencies, *Ecology and Society*, Vol.*3*, No.2, Art.8, 7 p.

Kababya, D., Perevolotsky, A., Bruckental, I. & Landau, S. (1998). Selection of diets by dual-purpose Mamber goats in Mediterranean woodland. *Journal of Agricultural Science*, *131*, 221-228.

Kelly, C. A. & Dyer, R. J. (2002). Demographic consequences of inflorescence-feeding insects for *Liatris cylindraceae*, an iteroparous perennial. *Oecologia*, *132*, 350-360.

Kesting, S., Wrage, N. & Isselstein, J. (2009). Herbage mass and nutritive value of herbage of extensively managed temperate grasslands along a gradient of shrub encroachment. *Grass Forage Science*, *64*, 246-254.

Köchy, M. & Wilson, S. D. (2000). Competitive effects of shrubs and grasses in prairie. *OIKOS*, *91*, 385–395.

Kronberg, S. L. & Walker, J. W. (2007). Learning Through Foraging Consequences: A Mechanism of Feeding Niche Separation in Sympatric Ruminants. *Rangelands Ecology & Management*, 60, 195–198.

Ksiksi, T. & Laca, E. A. (2000). Can social interactions affect food searching efficiency of cattle? *Rangeland Journal*, 22, 235–242.

Laca, E. A. (2009). New Approaches and tools for grazing management. *Rangelands Ecology & Management*, 62, 407-417.

Launchbaugh, K. (ed.) (2006). Targeted Grazing: A Natural Approach to Vegetation Management and Landscape Enhancement. Cottrell Printing, Centennial. p. 208.

Launchbaugh, K. L., Provenza, F. D. & Pfister, J. A. (2001). Herbivore response to anti-quality factors in forages. *Journal of Range Management*, 54, 431–440.

Leimu, R. & Lehtilä, K. (2006). Effects of multiple herbivores on the population dynamics of a perennial herb. *Basic and Applied Ecology*, 7, 224-235.

Magda, D., Chambon-Dubreuil, E., Agreil, C., Gleizes, B. & Jarry, M. (2009). Demographic analysis of a dominant shrub (*Cytisus scoparius*): Prospects for encroachment control. *Basic and Applied Ecology*, 10, 631-639.

Malkinson, D. & Kadmon, R. (2007). Vegetation dynamics along a disturbance gradient: Spatial and temporal perspectives. *Journal of Arid Environments*, 69, 127–143.

Meuret, M. (1988). Feasibility of in vivo digestibility trials with lactating goats browsing fresh leafy branches. *Small Ruminant Research*, 1, 273-290.

Meuret, M. (1997). Préhensibilité des aliments chez les petits ruminants sur parcours en landes et sous-bois. *INRA Prod. Anim.*, 10, 391-401.

Meuret, M. & Bruchou, C. (1994). Modélisation de l'ingestion selon la diversité des choix alimentaires réalisés par la chèvre au pâturage sur parcours. In: Institut de l'Elevage. 1ères Renc. Rech. Ruminants. Paris, France. p. 225-228.

Meuret, M., Débit, S., Agreil, C. & Osty, P. L. (2006). Eduquer ses veaux et génisses: un savoir empirique pertinent pour l'agroenvironnement en montagne. *Natures Sciences Sociétés*, 14, 343-352.

Meuret, M. & Giger-Reverdin, S. (1990). A comparison of two ways of expressing the voluntary intake of oak foliage-based diets in goats raised on rangelands. Reproduction, Nutrition, Développment 2(suppl.): 205.

Meuret, M., Viaux, C. & Chadoeuf, J. (1994). Land heterogeneity stimulates intake rate during grazing trips. *Annales de Zootechnie*, 43, 296-296.

Morley, F. H. W. (1981). Grazing animals. World animal science B1, Neimann-Sorensen A. & Tribe D.E. (Eds), Elsevier, UK, 411 p.

Moro, M. J., Pugnaire, F. I., Haase, P. & Puigdefabregas J. (1997). Mechanisms of interaction between a leguminous shrub and its understorey in a semi-arid environment. *ECOGRAPHY*, *20*, 175-184.

Papachristou, T., Dziba, L. & Provenza, F. (2005). Foraging ecology of goats and sheep on wooded rangelands. *Small Ruminant Research*, *59*, 141-156.

Papanastasis, V. P., Yiakoulaki, M. D., Decandia, M. & Dini-Papanastasi, O. (2008). Integrating woody species into livestock feeding in the Mediterranean areas of Europe. *Animal Feed Science and Technology*, *140*, 1-17.

Parsons, A. J. & Dumont, B. (2003). Spatial heterogeneity and grazing processes. *Anim. Res.*, *52*, 161–179.

Pataki, D. E., Billings, S. A., Naumburg, E. & Goedhart, C. M. (2008). Water sources and nitrogen relations of grasses and shrubs in phreatophytic communities of the Great Basin Desert. *Journal of Arid Environments*, *72*, 1581–1593.

Pihlgren, A. & Lennartsson, T. (2008). Shrub effects on herbs and grasses in semi-natural grasslands: positive, negative or neutral relationships? *Grass and Forage Science*, *63*, 9–21.

Pontes, L. da, S., Agreil, C., Magda, D., Gleizes, B. & Fritz, H. (2010). Feeding behaviour of sheep on shrubs in response to contrasting herbaceous cover in rangelands dominated by *Cytisus scoparius* L. *Applied Animal Behaviour Science*, *124*, 35-44.

Pontes, L., da, S., Magda, D., Jarry, M., Gleizes, B. & Agreil, C. (2012). Shrub encroachment control by browsing: targeting the right demographic process. *Acta Oecologica*, *45*, 25-30.

Provenza, F. (1995). Postingestive feedback as an elementary determinant of food preference and intake in ruminants. *Journal of Range Management*, *48*, 2-17.

Provenza, F., Villalba, J., Dziba, L., Atwood, S. & Banner, R. (2003). Linking herbivore experience, varied diets, and plant biochemical diversity. *Small Ruminant Research*, *49*, 257-274.

Rogosic, J., Estell, R., Ivankovic, S., Kezic, J. & Razov, J. (2008). Potential mechanisms to increase shrub intake and performance of small ruminants in Mediterranean shrubby ecosystems. *Small Ruminant Research*, *74*, 1-15.

Sardans, J., Peñuelas, J., Prieto, P. & Estiarte, M. (2008). Drought and warming induced changes in P and K concentration and accumulation in plant biomass and soil in a Mediterranean shrubland. *Plant Soil, 306,* 261–271.

Sih, A. & Christensen, B. (2001). Optimal diet theory: When does it work, and when and why does it fail? *Animal Behaviour, 61,* 379-390.

Slater, P. J. B. (1999). Essentials of animal behaviour. Cambridge University Press, Cambridge, UK. 244 p

Spalinger, D. & Hobbs, N. (1992). Mechanisms of foraging in mammalian herbivores: New models of functional response. *Am Nat, 140,* 325-348.

Stephens, D. W. & Krebs, J. R. (1986). Foraging Theory. Princeton University Press, Princeton, USA. 247 p.

Suding, K. N., Lavorel, S., Chapin, F. S., Cornelissen, J. H. C., Diaz, S., Garnier, E., Goldberg, D., Hooper, D., Jackson, S. T. & Navas, M. L. (2008). Scaling environmental change through the community-level: a trait-based response-and-effect framework for plants. *Global Change Biology, 14,* 1125-1140.

Tolera, A., Khazaal, K. & Orskov, E. R. (1997). Nutritive evaluation of some browse species. *Animal Feed Science and Technology, 67,* 181-195.

Valderrábano, J. & Torrano, L. (2000). The potential for using goats to control *Genista scorpius* shrubs in European black pine stands. *Forest Ecology and Management, 126,* 377-383.

Van Soest, P. J. (1994). Nutritional ecology of the ruminant (2nd edition). Cornell University press, New York, USA, 476 p.

Villalba, J. & Provenza, F. (2005). Foraging in chemically diverse environments: energy, protein, and alternative foods influence ingestion of plant secondary metabolites by lambs. *Journal of Chemical Ecology, 31,* 123-138.

Villalba, J. J., Soder, K. J. & Laca, E. A. (2009). Understanding diet selection in temperate biodiverse pasture systems. *Rangeland Ecology & Management, 62,* 387-388.

Westoby, M. (1999). A LHS strategy scheme in relation to grazing and fire. *In* D., Eldridge, D., Freudenberger, Editors. Proceedings of the VI[th] International Rangeland Congress, Queensland, Australia: *Australian Rangeland Society,* 893-896.

In: Herbaceous Plants
Editor: Florian Wallner

ISBN: 978-1-62618-729-0
© 2013 Nova Science Publishers, Inc.

*Chapter 2*

# Biosynthesis and Regulation of Tobacco Alkaloids

*Tsubasa Shoji*[*] *and Takashi Hashimoto*
Graduate School of Biological Sciences,
Nara Institute of Science and Technology, Japan

## Abstract

In *Nicotiana* plants, nicotine and related pyridine alkaloids, such as nornicotine, anabasine and anatabine, are synthesized in underground roots and then translocated via the xylem to aerial parts, where they are mainly stored in vacuoles as defensive toxins against herbivorous insects. A series of structural genes involved in the synthesis and transport of these alkaloids have been isolated, mostly based on their homology or expression profiles, and shown to be expressed in distinct types of root cells. Nevertheless, enzymes catalyzing the late synthetic steps, including ring coupling, have remained elusive. Jasmonate signals, in response to insect herivory, and cross-talking with auxin and ethylene, coordinately activate the nicotine pathway genes through a signaling cascade consisting of *Nicotiana* COI1, JAZs, and the bHLH transcription factor MYC2. Two genetic loci, *NIC1* and *NIC2*, mutant alleles of which have been used to breed tobacco cultivars with low nicotine content, specifically control multiple structural genes of the nicotine pathway. A

---

[*] Corresponding author: Tsubasa Shoji. E-mail: t-shouji@bs.naist.jp. Telephone: +81-743-72-5524. Fax: +81-743-72-5529. Address: 8916-5 Takayama, Ikoma, Nara 630-0101, Japan.

group of closely related *ERF* transcription factor genes are clustered at the *NIC2* locus and deleted in the *nic2* mutant. Jasmonate-inducible *NIC2*-locus ERFs and MYC2 directly up-regulate the transcription of the nicotine pathway genes, recognizing specific *cis*-elements in the promoters of their downstream target genes.

# Introduction

After Jean Nicot, the French ambassador to Portugal, sent tobacco seeds from Brazil to Paris in 1560, tobacco smoking readily became widespread around the world [1], largely due to the addictive nature of smoking. Government revenues of many nations rely partially on taxation of tobacco consumption and a huge tobacco industry has developed. Today, however, the markedly harmful effect of smoking on human health is well recognized [2] and various strategies to encourage smoking cessation are being tried [3].

Alkaloids are nitrogenous and mostly alkaline organic compounds, constituting a diverse group of secondary metabolites with over 12,000 chemical forms, and are found mainly in plants [4]. Their pharmacological activities make many alkaloids usable as poisons, narcotics, and medicines. Nicotine and related pyridine alkaloids, such as nornicotine, anabasine, and anatabines (Figure 1), are synthesized and accumulate in species of the genus *Nicotiana* including tobacco (*Nicotiana tabacum*), and their importance in tobacco products, such as cigars, cigarettes, pipe tobacco, and chewing tobacco, as major determinants of smoke quality and smoking addiction, is well known [5,6]. Nicotine acts on nicotinic acetylcholine receptors, which are widely distributed in the nervous system [7]. When inhaled, nicotine easily crosses the blood–brain barrier, and can reach the central nervous system through the bloodstream very rapidly: on average, it takes only seven seconds for inhaled nicotine to reach the brain. Nicotine increases the levels of several neurotransmitters. Like other addictive substances, such as cocaine and heroin, nicotine triggers the increased release of dopamine in the reward circuits of the brain that are responsible for euphoria and relaxation, establishing the firm dependence of a smoking habitat. Nicotine is highly toxic, with 30 to 60 mg a lethal dose for human adults. It was used as an effective insecticide, and its analogs such as imidacloprid are still in use.

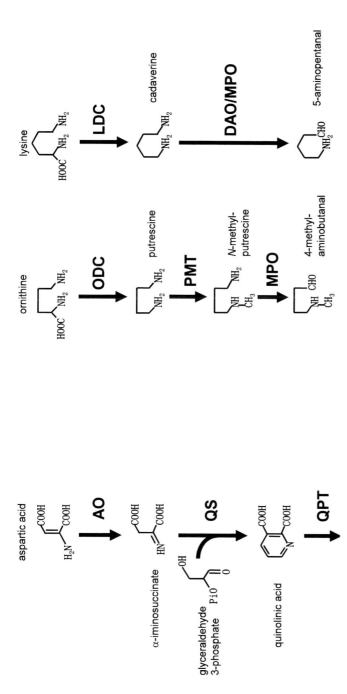

Figure 1. (Continued).

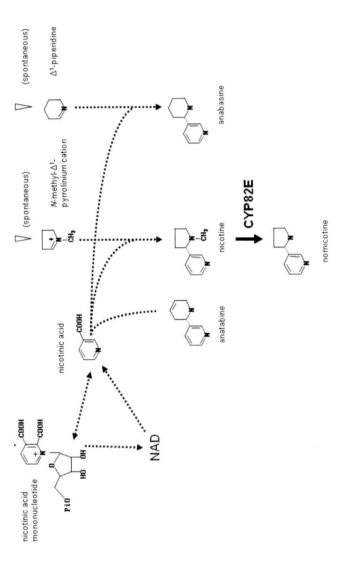

Figure 1. Biosynthetic pathway of nicotine and related pyridine alkaloids in *Nicotiana*. Solid arrows indicate defined biosynthetic steps catalyzed by known enzymes, whereas dashed arrows indicate multiple steps or ill-defined reactions. Details of how NAD and nicotinic acid are formed from nicotinic acid mononucleotide through cyclic routes are omitted. Enzyme abbreviations are as follows: ODC, ornithine decarboxylase; PMT, putrescine *N*-methyltransferase; MPO, *N*-methylputrescine oxidase; LDC, lysine decarboxylase; DAO, diamine oxidase; CYP82E, cytochrome P450 monooxygenase for nicotine *N*-demethylase; AO, aspartate oxidase; QS, quinolinate synthase; QPT, quinolinate phosphoribosyl transferase.

Nicotine was first isolated by the German chemists Posselt and Reimann in 1828, its chemical structure was determined by Pinner in 1893, and it was chemically synthesized by Pictet and Crépieux in 1904 [6]. Nicotine found in nature is an optically pure (-)-form, where the configuration at the C'-2 chiral center is (S). In tobacco, nicotine is the most abundant alkaloid, typically accounting for 90-95% of total alkaloid content, with the rest of the alkaloid pool primarily made up of secondary alkaloids, such as nornicotine, anabasine, and anatabine. The alkaloid profile within the genus *Nicotiana* is highly divergent, and in most species a single alkaloid is predominant [8]: *N. tomentosiformis* largely accumulates nornicotine, while in *N. glauca* (tree tobacco), anabasine is the major alkaloid.

## Biosynthesis

Alkaloids, usually derived from amino acids or their derivatives, can be classified based on their biogenic origins, and some of their biosynthetic pathways have been characterized extensively at the molecular level [9-14]. In tobacco, nicotine and related alkaloids are synthesized exclusively in roots and distributed throughout the plant via the xylem [15]. Roots of plantlets, root cultures, or cell cultures elicited for alkaloid production have been useful experimental materials, and the nicotine pathway has been studied through biochemical experiments followed by molecular cloning of the structural genes [16-20]. Genomic approaches to the study of *Nicotiana* species, such as EST and genome sequencing projects [TGI: http://www.tobaccogenome.org/; SGN: http://solgenomics.net/] [21, 22], are also advancing our understanding of the genes involved. The tobacco alkaloid biosynthetic pathway is shown in Figure 1.

### Pyrrolidine Formation

Nicotine is composed of two heterocyclic rings: a five-member pyrrolidine and a six-member pyridine ring. The pyrrolidine ring is derived from ornithine via the symmetric diamine putrescine. Putrescine is formed directly from ornithine by decarboxylation catalyzed by ornithine decarboxylase (ODC), which belongs to group IV of the pyridoxal 5'-phosphate-dependent decarboxylases [23-25]. Although putrescine is also derived from arginine through a pathway involving arginine decarboxylase, the contribution of this pathway to nicotine production is minor [26]. Putrescine is converted to higher polyamines, such as spermidine and spermine, or conjugated with other molecules such as cinnamate derivatives, in all higher plants [27].

At the first committed step in pyrrolidine formation, putrescine is converted to *N*-methylputrescine through *S*-adenosylmethionine-dependent *N*-methylation catalyzed by putrescine *N*-methyltransferase (PMT). Tobacco *PMT* expression was identified as being down-regulated in a low-nicotine mutant by subtraction between cDNA pools of the wild type and mutant [28]. *PMT* cDNA encodes a protein with sequence similarity to spermidine synthase (SPDS) [29], which transfers the aminopropyl moiety of decarboxylated *S*-adenosylmethionine to putrescine to generate spermidine, implying that *PMT* evolved from *SPDS* during the diversification of alkaloid-forming plants.

*N*-methylputrescine is oxidatively deaminated by *N*-methylputrescine oxidase (MPO) to 4-methylaminobutanal, which spontaneously cyclizes to *N*-methyl-$\Delta^1$-pyrrolinium cation. *MPO* cDNA encodes a protein similar to a group of diamine oxidases (DAOs), requiring copper and topaquinone as cofactors, that are widely distributed in nature [30, 31]. Recombinant MPO prefers *N*-methylputrescine to symmetric diamines, which are the preferred substrates of DAOs, indicating that MPOs arose from a DAO through optimization of the substrate specificity.

Down-regulation of *PMT* [32, 33] and *MPO* [34] reduces nicotine accumulation with concomitant increases of polyamines and other tobacco alkaloids lacking a pyrrolidine ring, such as anatabine and anabasine. In contrast to nicotine-rich roots, anatabine is the predominant alkaloid in tobacco BY-2 cultured cells elicited by jasmonates (JAs) [34, 35]. This anatabine-dominant profile is caused by a very low level of *MPO* expression. Accordingly, the overexpression of *MPO* in tobacco BY-2 cells restores the nicotine-accumulating profile [34].

## Pyridine Formation

The pyridine ring of nicotine is synthesized from nicotinic acid or its derivatives. Nicotinic acid is an intermediate in a pathway synthesizing nicotinamide adenine dinucleotide (NAD), a ubiquitous cofactor used in oxidoreduction reactions [36]. In dicotyledonous plants, the NAD pathway starts from aspartate. First, aspartate is oxidized by aspartate oxidase (AO) to form α-iminosuccinic acid. Next, α-iminosuccinic acid is condensed with glyceraldehyde-3-phosphate and cyclized by quinolinic acid synthase (QS), yielding quinolinic acid, which has a pyridine ring. The third reaction is the formation of nicotinic acid mononucleotide from quinolinic acid and phosphoribosyl pyrophosphate by quinolinic acid phosphoribosyl transferase (QPT) [37]. Nicotinic acid mononucleotide is converted to NAD and also to nicotinic acid in subsequent cyclic steps. The localization of *Arabidopsis* AO, QS, and QPT to plastids [38] suggests that the aspartate pathway is compartmentalized to the organelle. Tobacco *AO, QS,* and *QPT* genes are known to be regulated coordinately with nicotine biosynthesis, at least for certain gene family members that are highly expressed (see the case for *QPT2*, below); their transcripts are most abundant in nicotine-producing roots, but they are expressed at basal levels in leaves, and also subjected to regulation by *NIC* loci and activation by JA treatment [35, 37, 39].

## Coupling of the Pyrrolidine and Pyridine Rings

In contrast to the well defined early steps leading to the formation of pyridine and pyrrolidine rings, later steps involving the coupling of the two rings have remained unclear. Although an enzymatic activity of "nicotine synthase" catalyzing condensation in the presence of oxygen has been reported [40], the results have not been replicated.

Two orphan oxidoreductases of different families, A622 and berberine bridge enzyme-like protein (BBL), have been shown to be necessary for the late steps, although their exact enzymatic reactions have yet to be defined [41-43]. A622 belongs to the PIP family of NADPH-dependent oxiodoreductases, whose members include pinoresinol-laricresinol reductases, isoflavone reductases, and phenylcoumaran benzylic ether reductases [44]. BBL is a vacuole-localized FAD-containing oxidoreductase of the berberine bridge enzyme family that includes berberine bridge enzymes, carbohydrate oxidases, cannabinoid synthases and 6-hydroxynicotine oxidases [45]. *A622* and *BBL* were both isolated initially as genes repressed in a low-nicotine mutant [28, 43], and their expression profiles are closely associated with those of other genes involved in nicotine biosynthesis, in terms of tissue specificity, JA response, and genetic control by *NIC* loci [43, 46]. Suppression of *A622* or *BBL* by RNA interference severely reduced the formation of all of the pyridine alkaloids, with simultaneous over-accumulation of nicotinic acid *N*-glycoside, a detoxified form of free nicotinic acid, of *N*-methyl pyrrolinium cation for *A622* suppression [42], and of dihydromethanicotine, a novel nicotine metabolite for *BBL* suppression [43]. These metabolites were shown not to be accepted as substrates by recombinant A622 or BBL, but the possibility that they are derivatives of unstable substrates of the enzymes has not been excluded. A622 and BBL may synthesize a coupling-competent derivative of nicotinic acid, or may be directly involved in the coupling reaction itself.

## Nornicotine Formation

Nicotine is converted to nornicotine by nicotine *N*-demethylase (NND) [47], which belongs to the CYP82E subfamily of cytochrome P450 monooxygenases [48, 49]. In tobacco, at least three genes, *CYP82E4*, *E5*, and *E10*, encode functional NNDs, while a further two, *CYP82E2* and *E3*, encode inactive enzymes and are transcriptionally silent [48-51]. Nornicotine is typically a minor alkaloid in tobacco, representing about 3-5% of the total alkaloid content. In many tobacco populations, however, a few individuals known as "converters" can convert as much as 97% of the nicotine to nornicotine during leaf senescence and curing [52]. *CYP82E4* is the specific *NND* gene responsible for the conversion phenotype, which can arise when the normally silenced *CYP82E4* is reactivated in converter tobacco [48, 49, 53]. Knockout mutations in all the three functional *NND* genes, generated by chemical mutagenesis as well as transgenic suppression [48, 51, 54, 55], can drastically reduce accumulation of nornicotine to negligible levels. Since nornicotine is converted to a more harmful carcinogen, nitrosamine *N'*-nitrosonornicotine, during the curing and processing of harvested tobacco leaves [56], such reduction of nornicotine may be beneficial for tobacco breeding.

Natural variations in nornicotine formation found among *Nicotiana* wild species are attributable to genetic changes in *NND* genes in the following cases. The two progenitors of allotetraploid tobacco are *N. sylvestris* and *N. tomentosiformis* [57]. In contrast to nicotine-dominant tobacco, nornicotine accumulates substantially in senescing leaves of *N. sylvestris* and in both green and senescing leaves of *N. tomentosiformis*. Tobacco *CYP82E3* and *E4* genes are both derived from *N. tomentosiformis* [58], and *CYP82E2* from *N. sylvestris* [59]. Although tobacco *CYP82E2* and *E3* encode inactive NNDs and *CYP82E4* is transcriptionally silent, all ancestral orthologs of these genes *in N. sylvestris and N. tomentosiformis* encode active NNDs and are expressed in green and senescing leaves, accounting for the nornicotine accumulation pattern [58, 59]. Comparing sequences, stable mutations causing substitutions of functionally important amino acid residues were found in tobacco-inactive *CYP82E2* and *E3* [58, 59]. We can therefore infer that the ancestral *CYP82E* genes necessary for nornicotine accumulation acquired stable mutations in *CYP82E2* and *E3* and *an unstable mutation* causing silencing in *CYP82E4*, which collectively cause the inactivation of the tobacco genes. In another example, *N. langsdorffii* is totally devoid of nornicotine, whereas its closely related species *N. alata* accumulates this alkaloid abundantly. *N. alata* has four *CYP82E* genes encoding functional NNDs, whereas *N. langsdorffii* has two *CYP82E*s, both of which are inactivated by distinct mechanisms: one, encoding a functional enzyme, is not expressed at all, while the other is weakly expressed but contains a one-nucleotide deletion in the first exon, producing a non-functional truncated protein [60]. Genetic analysis based on interspecific crossing between *N. langsdorffii* and *N. alata* revealed that duplicated *CYP82E* genes in both species are genetically linked, and that the *N. alata CYP82E* locus is solely responsible for nornicotine accumulation [60].

## Anabasine and Anatabine Formation

Anabasine is composed of pyridine and piperidine rings. Like nicotine, nicotinic acid is incorporated into the pyridine ring in anabasine. On the other hand, the piperidine ring of anabasine, corresponding to the pyrrolidine ring of nicotine, is synthesized from lysine through decarboxylation by lysine decarboxylase (LDC) to produce cadaverine, and then oxidation by DAO/MPO to produce 5-aminopentanal, followed by spontaneous intramolecular cyclization resulting in $\Delta^1$-piperidine. The piperidine ring of anatabine is not derived from lysine; both rings of anatabine originate from nicotinic acids. The ring-coupling step of anabasine and anatabine remains unclear. Since the accumulation of all the pyridine alkaloids is decreased by *A622* or *BBL* suppression [41-43], the coupling step(s) involving A622 and BBL may be common to all tobacco alkaloids.

## Cell-specific Biosynthesis in Roots

Molecular localization of the gene transcripts and proteins involved in the nicotine pathway provides precise information about the cell types in which biosynthesis occurs. PMT and A622 have been thoroughly characterized by protein immunolocalization and promoter analyses, and shown to be expressed in the same cell types in the roots [46, 61]. Growing root tips often contain higher concentrations of nicotine than leaf lamina, and are considered the main site of biosynthesis. Consistent with this, PMT and A622 accumulate at higher levels in the apical parts of the roots. In the root apex just behind the meristematic region, both proteins are abundant throughout the endodermis and cortex. In the differentiated region of the roots where root hairs are present, the cortex, endodermis, and xylem display *PMT* and *A622* promoter activity, which is strongest in the outermost cortex layer, the endodermis layer, and parenchyma cells around xylem vessels. Gene expression in the xylem tissues may facilitate nicotine loading into the xylem stream. The expression patterns shown in detail for *PMT* and *A622* may apply to other genes of the pathway, as we have observed *QPT* [62] and *MATE1* [63] to be expressed in nearly identical cell types in the tissues examined. This observed coordination of multiple genes implies a shared regulatory mechanism conferring cell specificity.

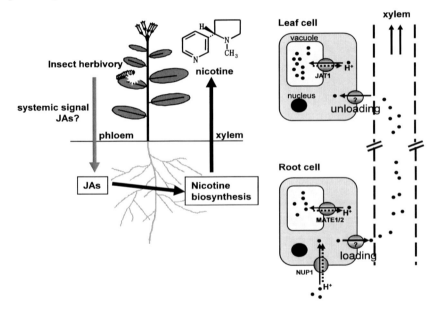

Figure 2. Nicotine transport and storage in tobacco. Insect herbivory in leaves induces nicotine production in roots by transmitting the JA signal systemically through phloem. After nicotine (represented by black dots in the right part of the diagram) is produced in root cells, it moves up via the xylem from roots to leaves. Putative membrane carriers, JAT1, MATE1/2, NUP1 and unknown proteins necessary for active transport of alkaloids, are presented schematically. Solid and dashed lines indicate the presumed directions of nicotine and proton movement, respectively, through these transporters.

# Transport

## Long-distance Transport from Roots to Leaves

Long-distance translocation of nicotine from roots to shoots was demonstrated by grafting between nicotine-producing tobacco and non-producing tomato; nicotine accumulated in tomato scions grafted on tobacco stocks [15]. Nicotine is produced exclusively in roots and then loaded into the xylem, a vasculature connecting roots and shoots, and easily moves up with the transpiration stream to leaves, where it plays defensive roles mainly against insect herbivores (Figure 2). Indeed, nicotine can be detected in tobacco xylem sap at high concentrations [64, 65]. To allow xylem loading and unloading, nicotine efflux from root cells and influx into leaf cells should be carried out at plasma membranes (Figure 2), but no nicotine transporters involved in these processes have yet been reported.

Although little is known about the control of inter-organ translocation of alkaloids, there is an intriguing case of natural variation, in which *N. alata* does not accumulate alkaloids in the leaf whereas the closely related species *N. langsdorffii* does. *N. alata* synthesizes alkaloids in the root, but lacks the ability to mobilize them to the aerial parts [65]. Interspecific grafting experiments between *N. alata* and *N. langsdorffii* indicate that roots of *N. alata* are unable to translocate alkaloids to their shoot system. Interestingly, genetic studies involving interspecific hybrids between *N. alata* and *N. langsdorffii* showed that the non-translocation phenotype is dominant over the translocation phenotype. These results indicate that a mechanism to retain tobacco alkaloids within the root organ has evolved in *N. alata*.

## Nicotine Transporters

To prevent cytotoxic effects at high concentrations, alkaloids are usually sequestered into the vacuole, a plant cell organelle typically used for bulk storage of harmful or waste chemicals. Nicotine, a weakly basic alkaloid, is uncharged and lipophilic under slightly alkaline conditions and can partially pass through the tonoplast by simple diffusion. Once in the acidic vacuole, however, nicotine can be protonated and trapped as a membrane-impermeable hydrophilic molecule. Such a scheme, termed the ion-trap mechanism, has been postulated for weakly basic compounds including nicotine. In addition to the ion-trap mechanism, active transport mediated by membrane-localized carriers also plays a major part in vacuolar sequestration.

Multidrug and toxic compound extrusion (MATE)-type transporters, JA-inducible alkaloid transporter 1 (JAT1), and the homologous protein pair MATE1 and MATE2 have been identified as tonoplast-localized nicotine transporters in tobacco (Figure 2) [63, 66]. The MATE transporters are one of five families comprising the multidrug transporter superfamily and have been shown to efflux low-molecular-weight compounds ranging from organic cations to metal ions as drug/H+ or drug/Na+ antiport systems [67]. Plants are particularly rich in MATE transporters, possibly reflecting their vast variety of secondary metabolites.

The MATE-type transporter *JAT1* was first identified through cDNA-amplified fragment length polymorphism-based profiling as a gene co-induced by methyl JA in tobacco BY2 cultures with other nicotine biosynthesis genes [35]. JAT1 shows high similarity to *Arabidopsis* DTX1, which mediates the efflux of xenobiotics at the plasma membrane [68]. It is expressed in leaves, stems, and roots, and is localized to the tonoplast in leaves [66]. Unlike other nicotine pathway genes, *JAT1* expression is not subject to regulation by *NIC* loci [39]. JAT1 showed nicotine transport activity when expressed in yeast, and further biochemical analysis using proteoliposomes reconstituted with JAT1 and the ΔpH-generating bacterial $F_0F_1$-ATPase demonstrated that JAT1 functions as a $H^+$-antiporter, transporting nicotine and other alkaloids but not flavonoids [66].

Tobacco MATE1 and MATE2 [63] are phylogenetically related to *Arabidopsis* TT12 [69] and tomato MTP77 [70], both of which have been implicated in the vacuolar sequestration of flavonoids. In contrast to *JAT1*, which is expressed in nearly every organ, *MATE* genes are specifically expressed in nicotine-producing root cells, and are regulated by *NIC* genes and JA in concert with other structural genes [63]. The localization of MATEs to tonoplasts has been shown by several experimental approaches including immunoelectron microscopy. Down-regulation of *MATE*s rendered tobacco roots more sensitive to exogenous application of nicotine, implying that MATEs are involved in the movement of nicotine *in planta*. In contrast, *MATE1* overexpression in cultured tobacco cells induced cytoplasmic acidification after either JA elicitation, which may trigger *de novo* nicotine production, or the addition of nicotine. When expressed in yeast cells, MATE1 reduced the amount of nicotine incorporated into the cells from the medium, to which it was added, indicating its nicotine transport activity. MATEs may be involved in the vacuolar sequestration of nicotine in nicotine-synthesizing root cells [63].

Other than the tonoplast-localized MATE-type transporters, tobacco nicotine uptake permease 1 (NUP1) has been reported to be a plasma membrane-localized nicotine transporter of the purine uptake permease (PUP) transporter family (Figure 2), whose founding members, *Arabidopsis* PUP1 and PUP2, are responsible for proton-mediated uptake of purines and cytokinins [71, 72]. The expression pattern of *NUP1* parallels those of other nicotine pathway genes in terms of *NIC* regulation, JA responsiveness, and tissue specificity, and indeed *NUP1* mRNA accumulates highly in root tips [71, 73]. In tobacco BY-2 cells, a NUP1-GFP fusion protein localizes to the plasma membrane. NUP1 expressed in yeast preferentially transports nicotine over other pyridine alkaloids. In hairy root cultures, suppression of *NUP1* by RNA interference decreases total nicotine levels in the tissues, but increases their accumulation in the culture medium. The reduced nicotine levels in hairy roots, which are also observed in the roots and leaves of plants regenerated from them, are caused by the down-regulation of nicotine biosynthesis, implying that *NUP1* affects nicotine metabolism through a mechanism that has yet to be determined [71].

# Regulation

## NIC Regulatory Genes

Genetic approaches using appropriate mutants are valuable for the elucidation of plant metabolic pathways and their regulation [74], as exemplified by studies on flavonoid biosynthesis that led to the identification of R- and C1-type transcriptional regulators. Contrary to mutants with visible phenotypes, such as plant pigment mutants, the isolation of alkaloid biosynthesis mutants is a challenging task and thus few such mutants have been reported [75, 76]. During the extensive history of tobacco cultivation, a naturally occurring mutant with low nicotine levels was occasionally found, initially in the early 1930s. The genetic basis of this mutant has been revealed through efforts to breed commercial tobacco varieties for low-nicotine cigarette production [77]; we now know that semi-dominant mutant alleles at the two unlinked loci *NIC1* and *NIC2* synergistically decrease nicotine concentration [77], and down-regulate the expression of multiple structural genes specific to the nicotine pathway, indicating the regulatory nature of *NIC* loci [28, 31, 39, 43, 63, 73]. Indeed, several structural genes involved in nicotine accumulation have been isolated after differential screening for tobacco genes that are specifically down-regulated in mutant roots, compared to wild-type roots, by cDNA subtraction [28], differential display [63, 73], and cDNA microarray [31, 39, 43]. Most metabolic and transport genes of the nicotine pathway (*ODC, PMT, MPO, AO, QS, QPT, A622, BBL, MATE1/2,* and *NUP1*) are categorized as *NIC*-controlled genes.

Microarray analysis has also provided a clue to the molecular identity of the *NIC2* locus; we found that the expression of a distinct set of tobacco ERF transcription factor genes was suppressed in the *nic2* mutant. At least seven *ERF* genes classified in AP2/ERF superfamily subgroup IXa are clustered at the *NIC2* locus, hereafter referred to as the *NIC2*-locus *ERF*s, and are found to be deleted altogether in the *nic2* mutant genome [39]. The deleted *ERF* genes all originate from the *N. tomentosiformis* ancestor, while the corresponding *ERF* genes of *N. sylvestris* origin appear to be intact in the *nic2* genome [39]. The presence of functionally redundant *ERF* genes at the *N. sylvestris*-derived locus likely explains why the total deletion of the *N. tomentosiformis*-derived *NIC2*-locus *ERF*s results in only a mild low-nicotine phenotype in the *nic2* mutant. The role of *NIC2*-locus ERFs in transcriptional regulation of nicotine pathway genes in the context of JA signaling is mentioned in the next section.

## Jasmonate

The chemical ecology of nicotine has been studied intensively using transgenic plants with altered nicotine content [78, 79]. Nicotine's role as a nectar repellent has also been revealed [80, 81]. In addition to basal production under normal conditions, nicotine is readily produced in response to attacks from insects [64]. Leaf damage caused by insect herbivory activates the wound response pathway, mediated mainly by JAs [82]. JA and its derivatives, collectively referred to as JAs, play signaling roles that are broadly associated with developmental regulation and defensive responses to stresses including herbivory and wounding [83, 84]. Reflecting possible defensive functions of natural products, elicitors that induce JA production and JA itself have been widely used to improve the productivity of useful phytochemicals in plant cell and tissue cultures [85-87]. The damage-induced formation of JAs in leaves and their translocation to roots via the phloem correlate with nicotine production in *Nicotiana* plants [82, 88], and JAs themselves may thus be the systematically transmitted chemical signal, although other possibilities cannot be excluded (Figure 2). In *Nicotiana* roots and cultured cells, JAs coordinate to activate most of the genes for nicotine biosynthesis and transport: *ODC, PMT, MPO, AO, QS, QPT, A622, BBL, MATE1/2, JAT1,* and *NUP1.*

JAs are synthesized from linolenic acid via the octadecanoid pathway, in which 13-lipoxygenase (LOX) catalyzes the oxygenation of linolenic acid at the first step [83]. The application of pharmacological inhibitors of the octadecanoid pathway [89] as well as transgenic silencing of *LOX* expression [90], both of which suppress JA production, can effectively reduce damage-induced production of nicotine. JA-isoleucine conjugate (JA-Ile) is the bioactive form of JA, and its formation is catalyzed by a JA-Ile-conjugating enzyme encoded by *JA Resistant 1 (JAR1)* in *Arabidopsis* [91] and by its orthologs *JAR4* and *JAR6* in *N. attenuata* [92]. Transgenic suppression of the conjugation step in *N. attenuata* by simultaneous silencing of *JAR4* and *JAR6* significantly reduced JA-induced nicotine accumulation, and this reduction was complemented by application of JA-Ile [92], supporting an essential role for JA-Ile in nicotine regulation.

In *Arabidopsis,* the molecular framework of JA signaling from perception to transcriptional activation has been elucidated mainly through molecular genetic approaches [84, 93]. Reception of the bioactive JA-Ile by the JA receptor coronatine insensitive1 (COI1), an F-box component of an SCF-type E3-ubiquitin ligase complex ($SCF^{COI1}$), can induce the recruitment of JA ZIM-domain (JAZ) proteins to the $SCF^{COI1}$ complex through a COI1-JAZ interaction, and JAZ proteins ubiquitinated by the ligase activity are removed by 26S proteasome-mediated degradation (Figure 3). There are 12 functionally redundant JAZs in *Arabidopsis,* and their primary structures are not highly conserved except for the Jas and TIFY motifs [94, 95], which are important interfaces for protein-protein interaction; JAZs form homo- and heterodimers by interacting at the TIFY motif, whereas the Jas motif mediates the COI1-JAZ interaction [96, 97]. JAZs act to connect various transcription factors with a complex containing the Groucho/Tup1-type co-repressor [98], which represses the transcription of nearby targeted genes, possibly through chromosomal remodeling. JA-dependent removal of JAZs releases JAZ-interacting transcription factors from the repressor complex, leading to the JA-responsive activation of downstream genes under the control of the liberated factors. Reflecting the pleiotropic aspects of the JA response, the list of JAZ target transcription factors is currently expanding: basic helix-loop-helix (bHLH)-family MYC2 and related MYC3 and MYC4, controlling a wide spectrum of JA-responsive genes [94, 95, 99, 100]; MYB-family members MYB21 and MYB24, required for stamen development [101]; and complex-forming bHLH- and R2R3 MYB-family factors, mediating anthocyanin accumulation and trichome initiation [102].

*Nicotiana* plants have functional counterparts of *Arabidopsis* COI1, JAZs, and MYC2, all of which are required for JA-responsive induction of nicotine biosynthesis (Figure 3) (see below for MYC2). In *Nicotiana*, suppression of *COI1* from *N. attenuata* [103] and tobacco [104] effectively impaired the JA- and wound-induced formation of nicotine as well as other typical JA responses. A number of *JAZ* genes have been identified in tobacco and *N. attenuata*, based on the presence of the Jas and TIFY motifs [104, 105]. Like their *Arabidopsis* counterparts, *Nicotiana JAZ*s were induced by JA [104, 105], while their proteins were degraded in response to JA, possibly by the ubiquitin-dependent proteasome system [104]. In tobacco, expression of JAZΔJas polypeptides - truncated forms of JAZs which lack the C-terminal Jas motifs that are important for function (involved in COI1-JAZ and JAZ-AtMYC2 interactions in *Arabidopsis*) but not for dimerization, and which act in a dominant-negative manner probably by forming non-functional JAZ dimers - also clearly inhibited JA induction of nicotine biosynthesis [104]. Interestingly, RNA interference suppression of *N. attenuata JAZh* significantly reduces nicotine levels, but enhances other subsets of JA-dependent responses [105]. These effects implicate NaJAZh as a repressor of most JA responses other than the nicotine response, whereas the contradictory reduction of the nicotine response may be explained by mutual cross-regulation of *JAZ*s, in which *JAZh* would negatively regulate other *JAZ*s that act as repressors of the nicotine response. The genes for nicotine biosynthesis are transcriptionally activated by JA. As analyzed in detail for *PMT* and *QPT* genes, JA-mediated induction of the nicotine pathway genes usually requires two distinct *cis*-elements, the G box and GCC box, in their proximal promoter regions (Figure 3) [61, 62, 106-109]. These elements are recognized by tobacco MYC2 and *NIC2*-locus ERF transcription factors, respectively [39, 62, 108, 110, 111]. The *MYC2* and *NIC2*-locus *ERF*s are transcriptionally induced by JA [39, 108, 111], and MYC2 and ERFs cooperatively *trans*-activate their target promoters when expressed transiently in tobacco cells [39, 62, 108, 111, 112]. Moreover, in transgenic *Nicotiana* plants and hairy roots, suppression of these transcription factor genes decreases the expression of genes involved in nicotine biosynthesis, consequently inhibiting the accumulation of tobacco alkaloids [39, 108, 110]. While MYC2 apparently controls a wide spectrum of JA-responsive defensive genes and is a target of JAZ repressors, directly interacting with them, *NIC2*-locus ERFs specifically regulate nicotine pathway genes and do not interact with JAZs in yeast two-hybrid assays [39, 108, 111].

Biosynthesis and Regulation of Tobacco Alkaloids 53

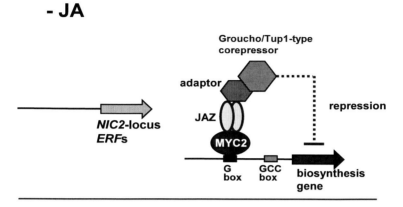

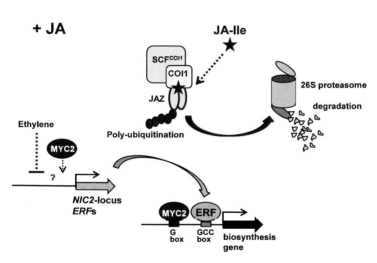

Figure 3. Model of jasmonate signaling leading to the transcriptional activation of nicotine biosynthesis genes. After JA-Ile is perceived at an interface formed between the COI1 subunit of the SCF$^{COI1}$ complex and recruited JAZs, JAZs are ubiquitinated by the SCF$^{COI1}$ complex and degraded by the 26S proteasome. The removal of JAZs in response to JA allows downstream MYC2 to activate the genes for nicotine biosynthesis, either directly by binding to their promoters at G boxes or indirectly through the activation of *NIC2*-locus *ERFs*. Together with MYC2, *NIC2*-locus ERF transcription factors, of which jasmonare-responsive transcription is under the control of MYC2, up-regulate all metabolic and transport genes of the nicotine pathway by binding at GCC boxes in their promoters. Ethylene signal may suppress the transcription of *NIC2*-locus *ERFs* to negatively regulate the JA response of nicotine biosynthesis. In the absence of JA signal input, interaction between JAZs and MYC2 recruits a complex containing the Groucho/Tup1-type co-repressor to biosynthesis gene promoters, which exerts a repressive effect on expression, possibly through chromatin remodeling.

*NIC2*-locus *ERFs* are members of the AP2/ERF superfamily subgroup IXa, which includes *Catharanthus roseus ORCA2* and *ORCA3* that are known to regulate the JA response of terpenoid indole alkaloid biosynthesis in this species [75, 113], suggesting that *ERFs* in the same subgroup are recruited to regulate distinct JA-responsive specialized metabolic pathways. All known *NIC*-controlled nicotine biosynthesis genes are immediate target genes of *NIC2*-locus *ERFs*. Indeed, at least eight GCC boxes are present in each of the examined *NIC*-controlled promoter regions, and their consensus sequence is 5'-A/CGCA/CNNCCA/T-3' [109], which is similar to but distinct from the canonical GCC box sequence, 5'-AGCCGCC-3', indicating a binding preference unique to *NIC2*-locus ERFs. Reflecting their functional conservation, the GCC boxes bound by tobacco *NIC2*-locus ERFs are also recognized by periwinkle ORCA3 [109].

Differential regulation of particular family members by *NIC2*-locus ERFs has been found for the *QPT* gene family, which encodes an enzyme involved in formation of the pyridine ring that contributes to both NAD and alkaloid synthesis. Two distinct genes, *QPT1* and *QPT2*, are conserved in *Nicotiana* species, but only the *QPT2* genes are controlled by *NIC2* locus-*ERFs*, being co-expressed with other nicotine biosynthesis genes, while *QPT1* expression is nearly constitutive in all examined tissues [62, 114]. As expected, promoters of *QPT2* genes harbor the GCC boxes recognized by *NIC2*-locus ERFs; in tobacco, three such GCC boxes are in the proximal promoter of *QPT2*, whereas none occur in that of *QPT1* [62]. These observations suggest the following scenario for the evolution of *Nictiana QPT* genes. After the two *QPTs* arose, possibly by duplication, the GCC boxes appeared in *QPT2* promoters, and thereby *QPT2* genes were recruited to the nicotine biosynthesis regulon controlled by *NIC2*-locus ERFs - allowing increased pyridine formation for nicotine synthesis - while the original role of *QPT* in sustaining the metabolic flow for NAD synthesis was retained by continuous *QPT1* expression [62].

While both MYC2 and *NIC2*-locus ERFs directly activate the transcription of nicotine pathway genes, is there hierarchical control among these transcription factors? When *MYC2* was silenced by RNA interference in tobacco hairy roots, all of the *NIC2*-locus *ERFs*, as well as nicotine biosynthesis and transporter genes, were down-regulated [108]. In contrast, *MYC2* expression level was unaffected either in the *nic2* mutant or in transgenic tobacco hairy roots expressing a dominant-repressive form of a *NIC2*-locus ERF [108]. These results suggest that *MYC2* regulates the *NIC2*-locus *ERF* genes, but not vice versa (Figure 3). Since MYC2-binding G boxes are predicted in the proximal promoter regions of several *NIC2*-locus *ERF* genes (our unpublished data), MYC2 may directly activate these *ERF* genes. Some *NIC2*-locus *ERF* genes are immediately induced by JA, while others are gradually induced at later time points [39]. Future studies should address how individual members of the *NIC2*-locus *ERF* gene family are regulated by MYC2, and possibly by particular ERFs themselves.

## Ethylene and Auxin

As anticipated from their overlapping roles in wound response and disease resistance, ethylene and JA interact synergistically or antagonistically in various signaling contexts. In nicotine biosynthesis, JA's activation of a number of structural genes, such as *ODC*, *PMT*, and *A622*, was effectively suppressed by simultaneous treatment with ethylene or its natural precursor, 1-aminocyclopropane-1-carboxylic acid (ACC), and the suppressive effect was abrogated when ethylene perception was blocked with specific inhibitors [115, 116]. The ethylene signal may integrate with JA signaling at the point around transcriptional regulation of *NIC2*-locus *ERF*s (Figure 3), since JA-mediated induction of the *ERF* transcripts is also suppressed by ACC treatment, even though the effects are not complete and are only apparent after several hours of treatment [39]. Ethylene has no significant influence on the expression of *MYC2*. The nicotine-tolerant herbivore *Manduca sexta* can induce production of ethylene, preventing an increase in nicotine synthesis [116-118]. An altered response to herbivores by an antagonistic interaction between ethylene and JA may ensure the re-allocation of resources from an ineffective nicotine-based defense.

Auxin, usually used in plant cell cultures at relatively high concentrations, is known to have a negative effect on alkaloid production, although whether it down-regulates the relevant gene expression directly or indirectly, by changing the state of differentiation of cells, is largely unclear. Low nicotine biosynthesis in cell and tissue cultures can be overcome by depleting auxin in the culture medium, which is accompanied by the concomitant activation of *ODC*, *PMT*, and *A622* [27, 28]. How auxin affects the expression of *NIC2*-locus *ERF*s and *MYC2* has yet to be addressed. To increase leaf mass as well as nicotine levels, removal of tobacco shoot apices, termed topping or decapitation, is a common pre-harvesting practice in tobacco cultivation. Topping may reduce the supply of auxin from young apices, and thus remove apical dominance for leaf growth and enhance nicotine production in roots, probably through gene activation that cancels the auxin-mediated down-regulation.

## Conclusion

Nicotine and related pyridine alkaloids are specialized bioactive metabolites that accumulate in *Nicotiana* as defensive toxins against insect attack. The alkaloids are produced in certain root cells and stored largely in vacuoles of every tissue in tobacco plants. A few nicotine transporters for active transport have been reported. Gene discovery efforts, including expression profiling of a low-nicotine mutant, have resulted in the isolation of many metabolic and transport genes involved in the alkaloid biosynthesis pathway, but molecular details of the late steps, including ring coupling, remain unclear. Our understanding of the regulation of the pathway by JA and other signals has been advanced by the identification of ERF and MYC2 transcription factors, which control a set of downstream structural genes mainly by binding to their promoters and are connected upstream with JA signaling components. Genetic manipulation of the pathway genes is beginning to allow us to alter the alkaloid profiles in the plants.

## References

[1] Goodman, J. (1993). *Tobacco in history: The cultures of dependence.* Abingdon, UK: Routledge.

[2] Hecht, S. S. (2003). Tobacco carcinogens, their biomarkers and tobacco-induced cancers. *Nat. Rev. Cancer*, 3, 733-744.

[3] Stead, L. F. & Lancaster, T. (2007). Interventions to reduce harm from continued tobacco use. *Cochrane Database Syst. Rev.*, 18, CD005231.

[4] Roberts, M. F. & Wink, M. (1998). *Alkaloids: Biochemistry, ecology, and medicinal applications.* New York, NY: Plenum Press.

[5] Davis, D. L. & Nielsen, M. T. (1999). *Tobacco: Production, chemistry and technology.* Oxford, UK: Blackwell Science.

[6] Gorrod, J. W. and Jacob, P. (1999). *Analytical determination of nicotine and related compounds and their metabolites.* Amsterdam, The Netherlands: Elsevier Science.

[7] Dome, P., Lazary, J., Kalapos, M. P. & Rihmer, Z. (2009). Smoking, nicotine and neuropsychiatric disorders. *Neurosci. Biobehav. Rev.*, 34, 295-342.
[8] Saitoh, F., Noma, M. & Kawashima, N. (1985). The alkaloid contents of sixty *Nicotiana* species. *Phytochemistry,* 24, 477-480.
[9] Hashimoto, T. & Yamada, Y. (1994). Alkaloiid biogenesis: molecular aspects. *Ann. Rev. Plant Physiol. Plant Mol Biol.*, 45, 257-285.
[10] Kutchan, T. M. (1995). Alkaloid biosynthesis: the basis for metabolic engineering of the medicinal plants. *Plant Cell*, 7, 1059-1070.
[11] de Luca, V. & Laflamme, P. (2001). The expanding universe of alkaloid biosynthesis. *Curr. Opin. Plant Biol.*, 4, 225-233.
[12] Facchini, P. J. (2001). Alkaloid biosynthesis in plants: biochemistry, cell biology, molecular regulation, and metabolic engineering applications. *Annu. Rev. Plant Physiol. Plant Mol. Biol.*, 52, 29-66.
[13] Hashimoto, T. & Yamada, Y. (2003). New genes in alkaloid metabolism and transport. *Curr. Opin. Biotechnol.*, 14, 163-168.
[14] Ziegler, J. & Facchini, P. J. (2008). Alkaloid biosynthesis: metabolism and trafficking. *Annu. Rev. Plant Biol.,* 59, 735-769.
[15] Dawson, R. F. (1942). Accumulation of nicotine in reciprocal grafts of tomato and tobacco. *Am. J. Bot.,* 29, 66-71.
[16] Leete, E. (1983). Biosynthesis and metabolism of the tobacco alkaloids. In S. W. Pelletier (ed.), *Alkaloids: Chemical and biological perspectives* (pp. 85-152) New York, NY: John Wiley & Sons.
[17] Bush, L. P. (1999). Alkaloid biosynthesis. In D. L. Davis, M. T. Nielsen (Eds.) *Tobacco: Production, chemistry and technology* (pp. 285-291) Oxford, UK: Blackwell Science.
[18] Bush, L., Hempfling, W. P. & Burton, H. (1999). Biosynthesis of nicotine and related compounds. In J. W. Gorrod, P. Jacob (Eds.) *Analytical determination of nicotine and related compounds and their metabolites* (pp. 13-44). Amsterdam, The Netherlands: Elsevier Science.
[19] Katoh, A., Oki, H., Inai, K. & Hashimoto, T. (2005). Molecular regulation of nicotine biosynthesis. *Plant Biotechnol.*, 22, 389-392.
[20] Shoji, T. & Hashimoto, T. (2011). Nicotine biosynthesis. In H. Ashihara, A. Crozier, A. Komamine (Eds.), *Plant metabolism and biotechnology* (pp. 191-216). Chichester, UK: John Wiley & Sons Ltd.
[21] Katoh, A., Yamaguchi, Y., Sano, H. & Hashimoto, T. (2003). Analysis of expression sequence tags from *Nicotiana sylvestris. Proc. Japan Acad. Ser. B.,* 79, 151-154.

[22] Rushton, P. J., Bokowiec, M. T., Laudeman, T. W., Brannock, J. F., Chen, X. & Timko, M. P. (2008). TOBFAC: The database of tobacco transcription factors. *BMC bioimformatics*, 9, 53.

[23] Sandmeier, E., Hale, T. I. & Christen, P. (1994). Multiple evolutionary origin of pyridoxal-5'-phosphate-dependent amino acid decarboxylases. *Eur. J. Biochem.*, 221, 997-1002.

[24] Michael, A. J., Furze, J. M., Rhodes, M. J. C. & Burtin, D. (1996). Molecular cloning and functional identification of a plant ornithine decarboxylase cDNA. *Biochem. J.*, 314, 241-248.

[25] Imanishi, S., Hashizume, K., Nakakita, M., Kojima, H., Matsubayashi, Y., Hashimoto, T., Sakagami, Y., Yamada, Y. & Nakamura, K. (1998). Differential induction of methyl jasmonate of genes encoding ornithine decarboxylase and other enzymes involved in nicotine biosynthesis in tobacco cell cultures. *Plant Mol. Biol.*, 38, 1101-1111.

[26] Chintapakorn Y. & Hamill J. D. (1990). Antisense-mediated reduction in ADC activity causes minor alterations in the alkaloid profile of cultured hairy roots and regenerated transgenic plants of *Nicotiana tabacum*. *Phytochemistry*, 68, 2465-2479.

[27] Kusano, T., Berberich, T., Tateda, C. & Takahashi, Y. (2007). Polyamines: essential factors for growth and survival. *Planta*, 228, 367-381.

[28] Hibi, N., Higashiguchi, S., Hashimoto, T. & Yamada, Y. (1994). Gene expression in tobacco low-nicotine mutants. *Plant Cell*, 6, 723-735.

[29] Hashimoto, T., Tamaki, K., Suzuki, K. & Yamada, Y. (1998). Molecular cloning of plant spermidine synthase. *Plant Cell Physiol.*, 39, 73-79.

[30] Heim, W. G., Sykes, K. A., Hildreth, S. B., Sun, J., Lu, R. H. & Jelesko, J. G. (2007). Cloning and characterization of a *Nicotiana tabacum* methylputrescine oxidase transcript. *Phytochemistry*, 68, 454-463.

[31] Katoh, A., Shoji, T. & Hashimoto, T. (2007). Molecular cloning of N-methylputrescine oxidase from tobacco. *Plant Cell Physiol.*, 48, 550-554.

[32] Sato, F., Hashimoto, T., Hachiya, A., Tamura, K. I., Chio, K. B., Morishige, T., Fujimoto, H. & Yamada, Y. (2001). Metabolic engineering of plant alkaloid biosynthesis. *Proc. Natl. Acad. Sci. USA*, 98, 367-372.

[33] Chintapakorn, Y. & Hamill, J. D. (2003). Antisense-mediated down regulation of putrescine *N*-methyltransferase activity in transgenic

*Nicotiana tabacum* L. can lead to elevated levels of anatabine at the expense of nicotine. *Plant Mol. Biol.*, 53, 87-105.

[34] Shoji, T. & Hashimoto, T. (2008). Why does anatabine, but not nicotine, accumulate in jasmonate-elicited cultured tobacco BY-2 cells. *Plant Cell Physiol.*, 49, 1209-1216.

[35] Goossens, A., Heakkinen, S.T., Laakso, I., Seppaenen-Laakso, T., Biondi, S., Sutter, V. D., Lammertyn, F., Nuutila, A. M., Soederlund, H., Zabeau, M., Inzé, D. & Oksman-Caldentey, K. M. (2003). A functional genomics approach toward the understanding of secondary metabolism in plant cells. *Proc. Natl. Acad. Sci. USA,* 100, 8595-8600.

[36] Katoh, A. & Hashimoto, T. (2004) Molecular biology of pyridine nucleotide and nicotine biosynthesis. *Front. Biosci.*, 9, 1577-1586.

[37] Sinclair, S. J., Murphy, K. J., Birch, C. D. & Hamill, D. (2000). Molecular characterization of quinolinate phosphoribosyltranferase (QPRTase) in *Nicotiana*. *Plant Mol. Biol.*, 44, 603-617.

[38] Katoh, A., Uenohara, K., Akita, M. & Hashimoto, T. (2006). Early steps in the biosynthesis of NAD in Arabidopsis start with asparatate and occur in the plastid. *Plant Physiol.*, 141, 851-857.

[39] Shoji, T., Kajikawa, M. & Hashimoto, T. (2010). Clustered transcription factor genes regulate nicotine biosynthesis. *Plant Cell*, 22, 3390-3409.

[40] Friesen, J. B. & Leete, E. (1990). Nicotine synthase- an enzyme from *Nicotiana* sprcies which catalyzes the formation of (S)-nicotine from nicotinic acid and 1-methyl-$\Delta^1$-pyrrolinium chloride. *Tetrahedron Lett.*, 31, 6295-6298.

[41] de Boer, K. D., Lye, J. C., Aitken, C. D., Su, A. K. & Hamill, J. D. (2009). The *A622* gene in *Nicotiana glauca* (tree tobacco): evidence for a functional role in pyridine alkaloid synthesis. *Plant Mol. Bol.*, 69, 299-312.

[42] Kajikawa, M., Hirai, N. & Hashimoto, T. (2009). A PIP-family protein is required for biosynthesis of tobacco alkaloids. *Plant Mol Biol.*, 69, 287-298.

[43] Kajikawa, M., Shoji, T., Katoh, A. & Hashimoto, T. (2011). Vacuole-localized berberine bridge enzyme-like proteins are required for a late step of nicotine biosynthesis in tobacco. *Plant Physiol.*, 155, 2010-2022.

[44] Gang, D. R., Kasahara, H., Xia, Z. Q., Vander Mijnsbrugge, K., Bauw, G., Boerjan, W., van Montagu, M., Davis, L. B. & Lewis, N. G. (1999). Evolution of plant defense mechanisms: relationships of phenylcoumaran benzylic ether reductases to pinoresinol-lariciresinol and isoflavone reductases. *J. Biol. Chem.*, 274, 7516-7527.

[45] Leferink, N. G., Heuts, D. P., Fraaije, M. W. & van Berkel, W. J. (2008). The growing VAO flavoprotein family. *Arch. Biochem. Biophys.,* 474, 292-301.
[46] Shoji, T., Winz, R., Iwase, T., Nakajima, K., Yamada, Y. & Hashimoto, T. (2002). Expression patterns of two tobacco isoflavone reductase-like genes and their possible roles in secondary metabolism in tobacco. *Plant Mol. Biol.,* 50, 427-440.
[47] Hao, D. Y. & Yeoman, M. M. (1996). Nicotine N-demethylase in cell-free preparations from tobacco cell cultures. *Phytochemistry,* 42, 325-329.
[48] Siminszky, B., Gavilano, L., Bowen, S. W. & Dewey, R. E. (2005). Conversion of nicotine to nornicotine in *Nicotiana tabacum* is mediated by CYP82E4, a cytochrome P450 monooxygenase. *Proc. Natl. Acad. Sci. USA,* 102, 14919-14924.
[49] Xu, D., Shen, Y., Chappell, J., Cui, M. & Nielsen, M. (2007). Biochemical and molecular characterization of nicotine demethylase in tobacco. *Physiol. Plantarum,* 125, 307-319.
[50] Gavilano, L. B. & Siminszky, B. (2007). Isolation and characterization of the cytochrome P450 gene *CYP82E5v2* that mediates nicotine to nornicotine conversion in the green leaves of tobacco. *Plant Cell Physiol.,* 48, 1567-1574.
[51] Lewis, R. S., Bowen, S. W., Keogh, M. R., & Dewey, R. E. (2010). Three nicotine demethylase genes mediate nornicotine biosynthesis in *Nicotiana tabacum* L.: Functional characterization of the *CYP82E10* gene. *Phytochemistry,* 71:1988-1998.
[52] Griffith, R. B, Valleau, W. D. & Stokes, G. W. (1955). Determination and inheritance of nicotine to nornicotine conversion in tobacco. *Science,* 121, 343-344.
[53] Chakrabarti, M., Bowen, S. W., Coleman, N. P., Meekins, K. M., Dewey, R. E. & Siminszky, B. (2008). *CYP82E4*-mediated nicotine to nornicotine conversion in tobacco is regulated by a senescence-specific signaling pathway. *Plant Mol. Biol.,* 66, 415-427.
[54] Gavilano, L. B., Coleman, N. P., Burnley, L. E., Bowman, M. L., Kalengamaliro, N. E., Hayes, A., Bush, L. & Siminszky, B. (2006). Genetic engineering of *Nicotiana tabacum* for reduced nornicotine content. *J. Agric. Food Chem.,* 54, 9071-9078.
[55] Lewis, R. S., Jack, A. M., Morris, J. W., Robert, V. J. M., Gavilano, L. B., Siminszky, B., Bush, L. P., Hayes, A. J. & Dewey, R. E. (2008). RNA interference (RNAi)-induced suppression of nicotine demethylase

activity reduces levels of a key carcinogen in cured tobacco leaves. *Plant Bioth. J.*, 6, 346-354.

[56] Bush, L. P., Cui, M., Shi, H. & Burton, H. R. (2001). Formation of tobacco specific nitrosamines in air-cured tobacco. *Rec. Adv. Tob. Sci.*, 27, 23-46.

[57] Clarkson, J. J., Lim, K. Y., Kovarik, A., Chase, M. W., Knapp, S. & Leitch, A.R. (2005). Long-term genome diploidization in allopolyploid *Nicotiana* section Repandae (Solanaceae). *New Physiol.*, 168, 241-252.

[58] Gavilano, L. B., Coleman, N. P., Bowman, M. L. & Siminszky, B. (2007). Functional analysis of nicotine demethylase genes reveals insights into the evolution of modern tobacco. *J. Biol. Chem.*, 282, 249-256.

[59] Chakrabarti, M., Meekins, K. M., Gavilano, L. B. & Siminszky, B. (2007). Inactivation of the cytochrome P450 gene *CYP82E2* by degenerative mutations was a key event in the evolution of the alkaloid profile of modern tobacco. *New Phytol.*, 175, 565-574.

[60] Pakdeechanuan, P., Teoh, S., Shoji, T. & Hashimoto, T. (2012). Nonfunctionalization of two *CYP82E* nicotine *N*-demethylase genes abolishes nornicotine formation in *Nicotiana langsdorffi*i. *Plant Cell Physiol.*, 53, 2038-2046.

[61] Shoji, T., Yamada, Y. & Hashimoto, T. (2000). Jasmonate induction of putrescine *N*-methyltransferase genes in roots of *Nicotiana sylvestris*. *Plant Cell Physiol.*, 41, 831-839.

[62] Shoji, T. & Hashimoto, T. (2011). Recruitment of a duplicated primary metabolism gene into the nicotine biosynthesis regulon in tobacco. *Plant J.*, 67, 949-959.

[63] Shoji, T., Inai, K., Yazaki, Y., Sato, Y., Takase, H., Shitan, N., Yazaki, K., Goto, Y., Toyooka, K., Matsuoka, K. & Hashimoto, T. (2009). Multidrug and toxic compound extrusion-type transporters implicated in vacuolar sequestration of nicotine in tobacco roots. *Plant Physiol.*, 149, 708-718.

[64] Baldwin, I. T. (1989). Mechanism of damage-induced alkaloid production in wild tobacco. *J. Chem. Ecol.*, 15, 1661-1680.

[65] Pakdeechanuan, P., Shoji, T. & Hashimoto, T. (2012). Root-to-shoot translocation of alkaloids is dominantly suppressed in *Nicotiana alata*. *Plant Cell Physiol.*, 53, 1247-1254.

[66] Morita, M., Shitan N., Sawada K., van Montagu M. C., Inzé D., Rischer H., Goossens A., Oksman-Caldentey K. M., Moriyama Y. & Yazaki K. (2009). Vacuolar transport of nicotine is mediated by a multidrug and

toxic compound extrusion (MATE) transporter in *Nicotiana tabacum*. *Proc. Natl. Acad. Sci. USA*, 106, 2447-2452.

[67] Omote, H., Hiasa, M., Matsumoto, T., Otsuka, M. & Moriyama, Y. (2006). The MATE proteins as fundamental transporters of metabolic and xenobiotic organic cations. *Trends Pharmacol. Sci.*, 27, 587-593.

[68] Li, L., He, Z., Girdhar, K. P., Tsuchiya, T. & Luan, S. (2002). Functional cloning and characterization of multidrug and heavy metal detoxification. *J. Biol. Chem.*, 277, 5360-5368.

[69] Marinova, K., Pourcel, L., Weder, B., Schwarz, M., Barron, D., Routaboul, J. M. & Klein, M. (2007). The Arabidopsis MATE transporter TT12 acts as a vacuolar flavonoid/$H^+$-antiporter active in proanthocyanidin-accumulating cells of the seed coat. *Plant Cell*, 19, 2023-2038.

[70] Mathews, H., Clendennen, S. K., Caldwell, C. G., Liu, X. L., Connors, K., Matheis, N., Schuster, D. K., Menasco, D. J., Wagoner, W., Lightner, J. & Wagner, D. R. (2007). Activation tagging in tomato identifies a transcriptional regulator of anthocyanin biosynthesis, modification, and transport. *Plant Cell*, 15, 1689-1703.

[71] Hildreth, S. B., Gehman, E. A., Yang, H., Lu, R. H., Ritesh, K. C., Harich, K. C., Yu, S., Lin, J., Sandoe, J. L., Okumoto, S., Murphy, A. S. & Jelesko, J. G. (2011). Tobacco nicotine uptake permease (NUP1) affects alkaloid metabolism. *Proc. Natl. Acad. Sci. USA*, 108, 18179-18184.

[72] Jelesko, J. G. (2012). An expanding role for purine uptake permease (PUP)-like transporters in plant secondary metabolism. *Front. Plant Physiol.*, 3, 78.

[73] Kidd, S. H., Melillo, A. A., Lu, R-H., Reed, D. G., Kuno, N., Uchida, K., Furuya, M. & Jelesko, J. G. (2006). The *A* and *B* loci in tobacco regulate a network of stress response genes, few of which are associated with nicotine biosynthesis. *Plant Mol. Biol.*, 60, 699-716.

[74] Memelink, J. (2005). The use of genetics to dissect plant secondary metabolism. *Curr. Opin. Plant Biol.*, 8, 230-235.

[75] van der Fits, L. & Memelink, J. (2000). ORCA3, a jasmonate-responsive transcriptional regulator of plant primary and secondary metabolism. *Science*, 289, 295-297.

[76] Millgate, A. G., Pogson, B. J., Wilson, I. W., Kutchan, T. M., Zenk, M. H., Berlach, W. L., Fist, A. J. & Larkin, P. J. (2004). Morphine-pathway block in top1 poppies. *Nature*, 431, 413-414.

[77] Legg, P. G. & Collins, G. B. (1971). Inheritance of percent total alkaloids in *Nicotiana tabacum* L. II. genetic effects of two loci in Burley21 X LA Burley21 populations. *Can. J. Genet. Cytol.*, 13, 287-291.

[78] Baldwin, I.T. (1998). Jasmonate-induced responses are costly but benefit plants under attack in native populations. *Proc. Natl. Acad. Sci. USA*, 95, 8113-8118.

[79] Steppuhn, A., Gase, K., Krock, B., Halitschke, R. & Baldwin, I. T. (2004). Nicotine's defensive function in nature. *PLoS Biol.*, 2, 1074-1080.

[80] Kessler, D. & Baldwin, I. T. (2006). Making sense of nectar scents: the effects of nectar secondary metabolites on floral visitors of *Nicotiana attenuata*. *Plant J.*, 49, 840-854.

[81] Kessler, D., Gase, K. & Baldwin, I. T. (2008). Field experiments with transformed plants reveal the sense of floral scents. *Science*, 321, 1200-1202.

[82] Baldwin, I. T., Schmelz, E. A. & Ohnmeiss, T. E. (1994). Wound-induced changes in root and shoot JA pools correlated with induced nicotine synthesis in *Nicotiana sylvestris* Spegazzini and Comes. *J. Chem. Ecol.*, 20, 2139-2157.

[83] Wasternack, C. (2007). Jasmonates: an update on biosynthesis, signal transduction and action in plant stress response, growth and development. *Ann. Bot.*, 100, 681-697.

[84] Browse, J. (2009). Jasmonate passes muster: a receptor and targets for the defense hormone. *Annu. Rev. Plant Biol.*, 60, 183-205.

[85] Gundlach, H., Müller, M. J., Kutchan, T. M. & Zenk, M. H. (1992). Jasmonic acid is a signal transducer in elicitor-induced plant cell cultutes. *Proc. Natl. Acad. USA*, 89, 2389-2393.

[86] Blechert, S., Brodschelm, W., Hölder, S., Kammerer, L. Kutchan, T.M., Mueller, M. J., Xia, Z. Q. & Zenk, M. H. (1995). The octadecanoic pathway: signal molecules for the regulation of secondary pathways. *Proc. Natl. Acad. Sci. USA*, 92, 4099-4105.

[87] Yukimune, Y., Tabata, H., Higashi, Y. & Hara, Y. (1996). Methyl jasmonate-induced overproduction of paclitaxel and baccatin III in *Taxus* cell suspension cultures. *Nat Biotechnol.*, 14, 1129-1132.

[88] Zhang, Z-P. & Baldwin, I. T. (1997). Transport of [2-$^{14}$C] jasmonic acid from leaves to roots mimics wound-induced changes in endogenous jasmonic acid pools in *Nicotiana sylvetris*. *Planta*, 203, 436-441.

[89] Baldwin, I. T., Schmelz, E. A. & Zhang, Z-P. (1996). Effects of octadecanoid metabolites and inhibitors on induced nicotine accumulation in *Nicotiana sylvestris*. *J. Chem. Ecol.*, 22, 61-74.

[90] Halitschke, R. & Baldwin, I. T. (2003). Antisense LOX expression increases herbivore performance by decreasing defense responses and inhibiting growth-related transcriptional reorganization in *Nicotiana attenuata*. *Plant J.*, 36, 794-807.

[91] Staswick, P. E. & Tiryaki, I. (2004). The oxylipin signal jasmonic acid is activated by an enzyme that conjugates it to isoleucine in Arabidopsis. *Plant Cell*, 16, 2117-2127.

[92] Wang, L, Allmann, S., Wu, J. & Baldwin, I. T. (2008). Comparisons of Lipoxygenase3- and Jasmonate-Resistant4/6-silenced plants reveal that jasmonic acid and jasmonic acid-amino acid conjugates play different roles in herbivore resistance of *Nicotiana attenuata*. *Plant Physiol.*, 146, 904-915.

[93] Chung, H. S., Niu, Y., Browse, J. & Howe, G. A. (2009). Top hits in contemporary JAZ: an update on jasmonate signaling. *Phytochemistry*, 70, 1547-1559.

[94] Chini, A., Fonseca, S., Fernández, G., Adie, B., Chico, J. M., Lorenzo, O., García-Casado, G., López-Vidriero, I., Lozano, F. M., Ponce, M. R., Micol, J. L. & Solano, R. (2007). The JAZ family of repressors is the missing link in jasmonate signalling. *Nature*, 448, 666-671.

[95] Thines, B., Katsir, L., Melotto, M., Niu, Y., Mandaokar, A., Liu, G., Nomura, K., He, S. Y., Howe, G. A. & Browse, J. (2007). JAZ repressor proteins are targets of the SCF$^{COI1}$ complex during jasmonate signalling. *Nature*, 448, 661-665.

[96] Melotto, M., Mecey, C., Niu, Y., Chung, H. S., Karsir, l., Yao, J., Zeng, W., Thines, B., Staswick, P., Browse, J., Howe, G. A. & He, S. Y. (2008). A critical role of two positively charged amino acids in the Jas motif of Arabidopsis JAZ proteins in mediating coronatine- and jasmonoyl isoleucine-dependent interactions with the COI1 F-box protein. *Plant J.*, 55, 979-988.

[97] Chini, A., Fonseca, S., Chico, J. M., Fernández-Calvo, P. & Solano, R. (2009). The ZIM domain mediates homo- and heteromeric interactions between *Arabidopsis* JAZ proteins. *Plant J.*, 59, 77-78.

[98] Pauwels, L., Barbero, G. F., Geerinck, J., Tilleman, S., Grunewald, W., Pérez, A. C., Chico, J. M., Bossche, R. V., Sewell, J., Gil, E., Garcia-Casado, G., Witters, E., Inzé, D., Long, J. A., de Jaeger, G., Solano, R.

& Goossens, A. (2010). NINJA connects the co-repressor TOPLESS to jasmonate signalling. *Nature,* 464, 788-791.

[99] Dombrecht, B., Xue, G. P., Sprague, S. J., Kirkegaard, J. A., Ross, J. J., Reid, J. B., Fitt, G. P., Sewelam, N., Schenk, P. M., Manners, J. M. & Kazan, K. (2007). MYC2 differentially modulates diverse jasmonate-dependent functions in *Arabidopsis. Plant Cell,* 19, 2225-2245.

[100] Fernández-Calvo, P., Chini, A., Fernández-Barbero, G., Chico, J.M., Gimenez-Ibanez, S., Geerinck, J., Eeckhout, D., Schweizer, F., Godoy, M., Franco-Zorrilla, J. M., Pauwels, l., Witters, E., Puga, M.I., Paz-Ares, J., Goossens, A., Reymond, P., de Jaeger, G. & Solano, R. (2011). The *Arabidopsis* bHLH transcription factors MYC3 and MYC4 are targets of JAZ repressors and act additively with MYC2 in the activation of jasmonate responses. *Plant Cell,* 23, 701-715.

[101] Song, S., Qi, T., Huang, H., Ren, Q., Wu, D., Chang, C., Peng, W., Liu, Y., Peng, J. & Xie, D. (2011). The Jasmonate-ZIM domain proteins interact with the R2R3-MYB transcription factors MYB21 and MYB24 to affect jasmonate-regulated stamen development. *Plant Cell,* 23, 1000-1013.

[102] Qi, T., Song, S., Ren, Q., Wu, D., Huang, H., Chen, Y., Fan, M., Peng, W., Ren, C. & Xie, D. (2011). The Jasmonate-ZIM-domain proteins interact with the WD-repeat/bHLH/MYB complexes to regulate jasmonate-mediated anthocyanin accumulation and trichome initiation in *Arabidopsis thaliana. Plant Cell,* 23, 1795-1814.

[103] Paschold, A., Halitschke, R. & Baldwin, I. T. (2007) Co(i)-ordinating defenses: NaCOI1 mediates herbivore-induced resistance in *Nicotiana attenuata* and reveals the role of herbivore movement in avoiding defenses. *Plant J.,* 51, 79-91.

[104] Shoji, T., Ogawa, T. & Hashimoto, T. (2008). Jasmonate-induced nicotine formation in tobacco is mediated by tobacco *COI1* and JAZ genes. *Plant Cell Physiol.,* 49, 1003-1012.

[105] Oh, Y., Baldwin, I. T. & Galis, I. (2012). NaJAZh regulates a subset of defense responses against herbivores and spontaneous leaf necrosis in *Nicotiana attenuata* plants. *Plant Physiol.,* 159, 769-788.

[106] Xu, B. & Timko, M. P. (2004). Methyl jasmonate induced expression of the tobacco putrescine *N*-methyltransferase genes requires both G-box and GCC-motif elements. *Plant Mol. Biol.,* 55, 743-761.

[107] Oki, H. & Hashimoto, T. (2004). Jasmonate-responsive regions in a *Nicotiana sylvestris PMT* gene involved in nicotine biosynthesis. *Plant Biotech.,* 21, 269-274.

[108] Shoji, T. & Hashimoto, T. (2011). Tobacco MYC2 regulates jasmonate-inducible nicotine biosynthesis genes directly and by way of the *NIC2*-locus *ERF* genes. *Plant Cell Physiol,*. 52, 1117-1130.

[109] Shoji, T. & Hashimoto, T. (2012). DNA-binding and transcriptional activation properties of tobacco *NIC2*-locus ERF189 and related transcription factors. *Plant Biotechnol.,* 29, 35-42.

[110] Todd, A. T., Liu, E., Polvi, S. L., Pammett, R. T. & Page, J. E. (2010). A functional genomic screen identifies diverse transcription factors that regulate alkaloid biosynthesis in *Nicotiana benthamiana*. *Plant J.,* 62, 589-600.

[111] de Boer, K., Tileman, S., Pauwels, L., Bossche, R.V., De Sutter, V., Vanderhaeghen, R., Hilson, P., Hamill, J. D. & Goossens, A. (2011). Apetala2/Ethyene Response Factor and basic helix-loop-helix transcription factors cooperatively mediate jasmonate-elicited nicotine biosynthesis. *Plant J.,* 66, 1053-1065.

[112] de Sutter, V., Vanderhaeghen, R., Tilleman, S., Lammertyn, F., Vanhoutte, I., Karimi, M., Inzé, D., Goossens, A. & Hilson, P. (2005). Exploration of jasmonate signaling via automated and standardized transient expression assays in tobacco cells. *Plant J.,* 44, 1065-1076.

[113] Shoji, T. & Hashimoto, T. (2012). Jasmonate-responsive transcription factors: new tools for metabolic engineering and gene discovery. In S. Chandra, H. Lata, A. Varma (Eds.), *Biotechnology for medicinal plants: Micropropagation and improvement.* (pp.345-357). Dordrecht, The Netherlands: Springer.

[114] Ryan, S. M., Cane, K. A., de Boer, K. D., Sinclair, S. J., Brimblecombe, R. & Hamill, J. D. (2012) Structure and expression of the quinolinate phosphoribosyltransferase (QPT) gene family in *Nicotiana*. *Plant Sci.,* 188-189, 102-110.

[115] Shoji, T., Nakajima, K. & Hashimoto, T. (2000). Ethylene suppresses jasmonate-induced gene expression in nicotine biosynthesis. *Plant Cell Physiol.,* 41, 1072-1076.

[116] Winz, R. A. & Baldwin, I. T. (2001). Molecular interactions between the specialist herbivore *Manduca sexta* (Lepidoptera, Sphingidea) and its natural host *Nicotiana attenuata*. IV. Insect-induced ethylene reduces jasmonate-induced nicotine accumulation by regulating putrescine *N*-methyltransferase transcripts. *Plant Physiol.,* 125, 2189-2202.

[117] Kahl, J., Siemens, D. H., Aerts, R. J., Gaebler, R., Kuehnemann, F., Preston, C. A. & Baldwin, I. T. (2000). Herbivore-induced ethylene

suppresses a direct defense but not a putative indirect defense against an adapted herbivore. *Planta,* 210, 336-342.
[118] von Dahl, C. C., Winz, R. A., Halitschke, R., Kühnemann, F., Gase, K. & Baldwin, I. T.(2007). Tuning the herbivore-induced ethylene burst: the role of transcript accumulation and ethylene perception in *Nicotiana attenuata. Plant J.,* 51, 293-307.

In: Herbaceous Plants
Editor: Florian Wallner

ISBN: 978-1-62618-729-0
© 2013 Nova Science Publishers, Inc.

*Chapter 3*

# Hybrid Lethality in *Nicotiana*: A Review with Special Attention to Interspecific Crosses between Species in Sect. *Suaveolentes* and *N. Tabacum*

*Takahiro Tezuka*[*]

Graduate School of Life and Environmental Sciences,
Osaka Prefecture University, Sakai, Osaka, Japan

## Abstract

Inviability of hybrids, often referred to as hybrid lethality, is a type of reproductive isolating mechanism. Hybrid lethality is observed in a number of plant species, including *Nicotiana* species, and can be an obstacle to the introduction of desirable genes into cultivated species by wide hybridization. In this chapter, I review hybrid lethality in *Nicotiana* with special attention to interspecific crosses between species in *Nicotiana* sect. *Suaveolentes* and *N. tabacum*. Most wild species in sect. *Suaveolentes* (which consists of species restricted to Australasia and Africa) yield inviable hybrids after crosses with the cultivated species *N. tabacum* ($2n = 48$, SSTT). Genetic studies have revealed that hybrid

---

[*] Corresponding author E-mail: tezuka@plant.osakafu-u.ac.jp.

lethality in *N. tabacum* × *N. debneyi* is caused by interaction between one or more genes on the Q chromosome of *N. tabacum* and a single dominant gene in *N. debneyi*. Gene(s) on the *N. tabacum* Q chromosome are also responsible for hybrid lethality in crosses involving most other *Suaveolentes* species. Most notably, genes from both S and T subgenomes of *N. tabacum* are responsible for hybrid lethality in *N. tabacum* × *N. occidentalis*. In addition, two species, *N. benthamiana* and *N. fragrans*, produced 100% viable hybrids from crosses with *N. tabacum*. These results provide a framework for discussing evolutionary processes leading to hybrid lethality in sect. *Suaveolentes*.

# Introduction

Reproductive isolation involves a variety of premating, postmating-prezygotic, and postzygotic isolating barriers in animals and plants (Stebbins, 1966; Coyne and Orr, 2004; Rieseberg and Blackman, 2010). Although these barriers contribute to species formation, they are obstacles for plant breeders, especially in breeding programs involving wide hybridization, where postmating-prezygotic and postzygotic isolating barriers are major problems. Over the past 80 years, extensive research has been carried out on these barriers using interspecific crosses in the genus *Nicotiana*. Several methods have been developed to bypass these barriers, although the mechanisms behind these barriers and the reasons for efficacy of the developed methods are not necessarily clear. In this chapter, I first review isolating barriers in *Nicotiana*, and then focus on hybrid lethality, primarily in crosses between species in sect. *Suaveolentes* and *N. tabacum*.

# Crossing Barriers in *Nicotiana*

The genus *Nicotiana* (Solanaceae) includes 76 species classified into 13 sections predominantly distributed in the Americas and Australia (Knapp et al., 2004). Numerous intrageneric crosses have been attempted with varying degrees of success by many researchers, with some failures attributable to postmating-prezygotic and postzygotic isolating barriers. Results of interspecific crosses in *Nicotiana* were well documented by Kostoff (1930) and McCray (1933). Their classifications are summarized as follows:

1 Crosses in which the pollen tubes do not reach the ovary and the flower falls (pollen-pistil incompatibility or incongruity).
2 Crosses in which the ovary is stimulated but without production of a hybrid embryo.
   a Stimulation stops with the induction of some cell divisions and growth in the nucellus.
   b Parthenocarpic seeds are produced.
   c Parthenogenetic (maternal) seeds are produced, either haploid or diploid.
3 Crosses in which true hybrid embryos are produced, but either no seed is formed or non-germinating seeds are produced.
   a Embryos die at about the four-cell stage.
   b Many cells are produced over a time period equivalent to one fourth or more of the usual period of embryonic development, although possibly without any appreciable growth in size.
   c Embryos apparently complete their growth, but most seeds are unable to germinate.
4 Crosses in which the seeds germinate, but the hybrid plants die in the seedling stage (hybrid lethality).
5 Crosses in which mature hybrid plants are obtained.
   a Dwarf plants are generated that remain small but otherwise mature and produce flowers (hybrid dwarfness).
   b Vigorous, healthy but sterile plants are produced (hybrid sterility).
   c Vigorous, healthy, and fertile or partially fertile plants are produced.

The above-mentioned isolating barriers have been observed independently or in combination with other barriers in various cross combinations.

Several methods for bypassing isolating barriers have been reported in *Nicotiana*. One method exploits the unilateral nature of pollen-pistil incompatibility (unilateral incompatibility) in *Nicotiana* (Christoff, 1928; Kostoff, 1930; McCray, 1932; East, 1935; Swaminathan, 1957; Tanaka, 1961; DeVerna et al., 1987; Kuboyama et al., 1994; Kuboyama and Takeda, 2000; Tezuka et al., 2007, 2010; Laskowska and Berbeć, 2012; Tezuka and Marubashi, 2012). When cross incompatibility is observed in one cross direction, an attempted reciprocal cross may be employed. If pollen-pistil incompatibility must be bypassed to obtain hybrid embryos or seedlings, bud pollination (Kuboyama et al., 1994), cut-style pollination (Swaminathan,

1957), mixed pollination (Kincaid, 1949), pollination using irradiation-killed or frozen-and-thawed-killed compatible pollen (mentor pollen) together with incompatible pollen (Pandey, 1977), and test-tube pollination (test-tube fertilization) (Ternovskii et al., 1976; Marubashi and Nakajima, 1985; DeVerna et al., 1987; Marubashi and Onosato, 2002; Tezuka and Marubashi, 2006a, 2012; Tezuka et al., 2007, 2010) may be successful. In addition, Swaminathan and Murty (1959) succeeded in overcoming cross incompatibility by using incompatible pollen exposed to X-rays for pollination in the incompatible cross.

When hybrid embryos are produced, but either no seed or non-germinating seeds are produced, ovule culture is frequently used (Reed and Collins, 1978; Shizukuda and Nakajima, 1982; Subhashini et al., 1985, 1986; Iwai et al., 1986; Chung et al., 1988; Tezuka et al., 2010). Reed and Collins (1980) reported that in some crosses, embryo abortion occurs as a result of endosperm malfunction; in these cases, embryo culture medium can substitute for the endosperm in providing nutrition to the developing hybrid embryo. Another issue, hybrid sterility, can be overcome by chromosome doubling or by increasing the number of chromosomes using colchicine treatment (Ternovskii et al., 1972, 1976; Lloyd, 1975; Subhashini et al., 1986; Chung et al., 1988; Trojak-Goluch and Berbeć, 2007) or tissue culture (DeVerna et al., 1987; Nikova and Zagorska, 1990; Nikova et al., 1991, 1999).

Other methods exist for bypassing postmating-prezygotic and postzygotic isolating barriers (including hybrid lethality). One such technique is interspecific bridge crossing. For example, although hybrid seedlings in reciprocal crosses between *N. repanda* and *N. tabacum* cannot be produced by conventional cross-pollination because of several barriers (unilateral incompatibility, seed abortion, and hybrid lethality), disease resistance can be transferred from *N. repanda* to *N. tabacum* using *N. sylvestris* as a bridging species, in combination with colchicine treatment (Stavely et al., 1973). In addition, direct hybridization of *N. repanda* with *N. tabacum* has been accomplished when *N. repanda* (4n) obtained by colchicine treatment is used as a parent (Pittarelli and Stavely, 1975). Somatic hybridization by protoplast fusion is also useful to obtain somatic hybrids or cybrids when crosses are difficult owing to isolating barriers (Nagao, 1978, 1979, 1982; Kumashiro and Kubo, 1986a; Kumashiro et al., 1988; Ilcheva et al., 2001; Fitter et al., 2005; Sun et al., 2005).

## Hybrid Lethality in *Nicotiana*

Hybrid plants from normal parents sometimes show weak growth or die before maturity. Several terms—hybrid lethality, hybrid weakness, hybrid necrosis, and hybrid inviability—have been used to describe this phenomenon. Hybrid lethality is observed in certain cross combinations in many plant species (Bomblies and Weigel, 2007), including many interspecific crosses in *Nicotiana*. These include crosses between cultivated species *N. tabacum* and *N. rustica*, between cultivated and wild species, and between wild species. I have summarized these cross combinations elsewhere (Tezuka, 2012).

In tobacco breeding, it is important to overcome hybrid lethality, as this mechanism, as well as other isolating barriers, can impede introduction of desirable genes into cultivated tobacco. After many attempts by different researchers, several practical methods to overcome or suppress hybrid lethality have been developed in *Nicotiana*.

Temperatures are critical for control of hybrid lethality. Hybrid seedlings show hybrid lethality at 28°C, a temperature normally suitable for growth of tobacco plants. When hybrid seedlings are cultivated at elevated temperatures generally ranging from 32–38°C, they may grow normally without exhibiting lethal symptoms and even flower (Manabe et al., 1989; Yamada et al., 1999; Tezuka and Marubashi, 2006a). Hybrid seedlings must be cultivated continuously from germination to maturity at elevated temperatures; however, if transferred from an elevated temperature to one below 28°C, they die.

Hybrid lethality can be effectively overcome through *in vitro* culture of seeds (Nikova and Zagorska, 1990) or explants, including germinated seeds (Ternovskii et al., 1972), cotyledons (Lloyd, 1975; Ternovskii et al., 1976; DeVerna et al., 1987; Yamada et al., 1999; Laskowska and Berbeć, 2012) and small leaves (Iwai et al., 1985), in the presence of appropriate plant growth regulators. Viable hybrids can be also obtained by application of the exogenous plant growth regulators auxin (Zhou et al., 1991) and cytokinin (Inoue et al., 1994, 1997; Yamada et al., 1999) to hybrid seedlings. Use of pollen or egg cells (ovules) exposed to γ-ray (Shintaku et al., 1988, 1989; Kitamura et al., 2003) or ion beam (Kitamura et al., 2003) irradiation for lethal crosses is also a successful technique for overcoming hybrid lethality. For more details of methods used to overcome or suppress hybrid lethality, see Tezuka (2012). While several practical methods, as mentioned above, have been developed, research addressing the mechanism of hybrid lethality only began in the past decade or so. Inviable hybrid seedlings in *Nicotiana* show several specific initial phenotypes (surface symptoms) depending on cross combinations (Yamada et al., 1999; Tezuka and Marubashi, 2012).

Hybrid lethality in this genus can be classified into five categories based on external phenotypes:

- Type I: browning of shoot apex and root tips.
- Type II: browning of hypocotyl and roots.
- Type III: yellowing of true leaves.
- Type IV: formation of multiple shoots.
- Type V: fading of shoot color.

The effectiveness of three methods to rescue inviable hybrids—cultivation at elevated temperatures, cotyledon culture, and cytokinin treatment—depends on type of hybrid lethality, at least with Types I-IV.

Different physiological processes are therefore thought to be associated with different lethality types (Yamada et al., 1999), and it is possible that different causal factors control different types of hybrid lethality.

## Hybrid Lethality in Crosses between *Suaveolentes* Species and *N. Tabacum*

In *Nicotiana*, hybrid lethality has been extensively studied in interspecific crosses between species in sect. *Suaveolentes* and *N. tabacum*. Section *Suaveolentes* includes 26 species, most of which are endemic to Australia, although three species are found in other locations. In particular, *N. debneyi* is distributed in eastern Australia, New Caledonia, and Lord Howe Island, while *N. fragrans* is distributed in New Caledonia and the Loyalty, Tonga, and Marquesas Islands (Goodspeed, 1954; Ladiges et al., 2011). The third non-Australian species, *N. africana*, is restricted to Namibia (southwest Africa), and represents the only *Nicotiana* species discovered to date in Africa (Merxmüller and Buttler, 1975). Species in sect. *Suaveolentes* are geographically isolated from the majority of species in other sections, which are found in the Americas.

All species in sect. *Suaveolentes* are allotetraploids and comprise an almost complete aneuploid series of $n = 16-24$, with only $n = 17$ as yet undiscovered. Section *Suaveolentes* is thought to have originated from a single polyploid event approximately 10 million years ago, followed by speciation to produce the species known today (Leitch et al., 2008). This hypothesis is supported by recent studies indicating that sect. *Suaveolentes* is monophyletic, based on molecular data from the internal transcribed spacer (ITS) region (Chase et al., 2003), plastid genes (Clarkson et al., 2004), and the nuclear-encoded chloroplast-expressed glutamine synthetase (ncpGS) gene (Clarkson et al., 2010). In addition, *Suaveolentes* species are genetically distant from *N. tabacum* (Chase et al., 2003; Clarkson et al., 2004, 2010). Although *Suaveolentes* species are valuable as sources of disease resistance (Burk and Heggestad, 1966; Smith, 1968; Bai et al., 1995; Li et al., 2006) and cytoplasmic male sterility (Smith, 1968; Kumashiro and Kubo, 1986a, 1986b; Kubo et al., 1988; Kumashiro et al., 1988; Nikova and Zagorska, 1990; Bonnett et al., 1991; Nikova et al., 1991; Fitter et al., 2005), the occurrence of hybrid lethality represents a barrier to introduction of desirable characteristics into *N. tabacum* through interspecific crosses involving this section.

To my knowledge, 22 species in sect. *Suaveolentes* have been crossed with *N. tabacum* and the resulting hybrid seedling viability reported. Results of these crosses are summarized in Table 1. Twenty *Suaveolentes* species show hybrid lethality after crosses with *N. tabacum*. In most cases, the hybrid lethality is of Type II (Yamada et al., 1999; Mino et al., 2002; Tezuka and Marubashi, 2004, 2006a; Tezuka et al., 2006, 2010, 2012; Iizuka et al., 2010; Laskowska and Berbeć, 2012). Type II hybrid lethality in *N. suaveolens* × *N. tabacum* is shown in Figure 1.

Only hybrid seedlings between *N. occidentalis* and *N. tabacum* show Type V lethality (Tezuka and Marubashi, 2012). Lethality of Types II and V is observed in these crosses at or below 28°C, but is completely suppressed at temperatures ranging from 34 to 37°C (Manabe et al., 1989; Marubashi and Kobayashi, 2002; Mino et al., 2002; Tezuka and Marubashi, 2004, 2006a, 2012; Tezuka et al., 2006, 2010, 2012; Iizuka et al., 2010). In contrast, two *Suaveolentes* species, *N. benthamiana* and *N. fragrans*, do not show hybrid lethality and produce 100% viable hybrids in crosses with *N. tabacum* (DeVerna et al., 1987; Iizuka et al., 2010, 2012; Tezuka et al., 2010).

# Causal Genes of Most Cases of Hybrid Lethality

When reciprocal crosses with *N. tabacum* were carried out using 19 of the above-mentioned *Suaveolentes* species producing inviable hybrids (i.e., all except for *N. wuttkei*), hybrid lethality was observed regardless of cross direction.

**Table 1. Hybrid lethality observed in crosses between *Suaveolentes* species and *N. tabacum***

| Suaveolentes species | Haploid chromosome number | $F_1$ phenotype | Suppression at elevated temperatures[a] | Factors responsible for hybrid lethality[a] | | References[b] |
|---|---|---|---|---|---|---|
| | | | | In *N. tabacum* | In *Suaveolentes* species | |
| *N. africana* | 23 | Type II lethality | Possible | Q chromosome | ND | 11, 12 |
| *N. amplexicaulis* | 18 | Type II lethality | Possible | Q chromosome | ND | 9, 11 |
| *N. benthamiana* | 19 | Viable | – | – | *hla1-2* | 1, 11, 14 |
| *N. cavicola* | 23 | Type II lethality | Possible | ND | ND | 11 |
| *N. debneyi* | 24 | Type II lethality | Possible | Q chromosome | *Hla1-1* | 3, 4, 7, 10, 11 |
| *N. excelsior* | 19 | Type II lethality | Possible | Q chromosome | ND | 11, 12 |
| *N. exigua* | 16 | Type II lethality | Possible | ND | ND | 11 |
| *N. fragrans* | 24 | Viable | – | – | *hla1-2* | 12 |
| *N. goodspeedii* | 20 | Type II lethality | Possible | Q chromosome | ND | 11, 12 |
| *N. gossei* | 18 | Type II lethality | Possible | Q chromosome | ND | 5, 11, 12 |
| *N. hesperis* | 21 | Type II lethality | Possible | ND | ND | 11 |
| *N. ingulba* | 20 | Type II lethality | Possible | Q chromosome | ND | 11, 17 |
| *N. maritima* | 16 | Type II lethality | Possible | Q chromosome | ND | 12 |
| *N. megalosiphon* | 20 | Type II lethality | Possible | Q chromosome | ND | 11, 12 |
| *N. occidentalis* | 21 | Type V lethality | Possible | S and T subgenomes | ND | 16 |
| *N. rosulata* | 20 | Type II lethality | Possible | ND | ND | 11 |
| *N. rotundifolia* | 22 | Type II lethality | Possible | ND | ND | 11 |

| Suaveolentes species | Haploid chromosome number | F₁ phenotype | Suppression at elevated temperatures[a] | Factors responsible for hybrid lethality[a] | | References[b] |
|---|---|---|---|---|---|---|
| | | | | In *N. tabacum* | In *Suaveolentes* species | |
| *N. simulans* | 20 | Type II lethality | Possible | Q chromosome | ND | 11, 13 |
| *N. suaveolens* | 16 | Type II lethality | Possible | Q chromosome | ND | 2, 3, 6, 8, 11 |
| *N. umbratica* | 23 | Type II lethality | Possible | ND | ND | 11 |
| *N. velutina* | 16 | Type II lethality | Possible | Q chromosome | ND | 12 |
| *N. wuttkei* | 16 | Type II lethality | ND | ND | ND | 15 |

[a]ND, not determined.
[b]1, DeVerna et al. (1987); 2, Manabe et al. (1989); 3, Yamada et al. (1999); 4, Marubashi and Kobayashi (2002); 5, Mino et al. (2002); 6, Tezuka and Marubashi (2004); 7, Tezuka and Marubashi (2006a); 8, Tezuka and Marubashi (2006b); 9, Tezuka et al. (2006); 10, Tezuka et al. (2007); 11, Iizuka et al. (2010); 12, Tezuka et al. (2010); 13, Tezuka et al. (2011); 14, Iizuka et al. (2012); 15, Laskowska and Berbeć (2012); 16, Tezuka and Marubashi (2012); 17, Tezuka et al. (2012).

This indicates that hybrid lethality is due to the interaction of coexisting heterologous genomes, and not to a cytoplasmic effect; i.e., both *N. tabacum* and *Suaveolentes* species possess causal genes encoded in nuclear genomes (Tezuka and Marubashi, 2004, 2006a, 2012; Tezuka et al., 2006, 2010, 2011, 2012; Iizuka et al., 2010).

To reveal genes causing hybrid lethality in *N. tabacum*, genetic analyses have been conducted using progenitors and monosomic lines of *N. tabacum*. *Nicotiana tabacum* ($2n = 48$, SSTT) is a natural allotetraploid (amphidiploid) derived from hybridization between *N. sylvestris* ($2n = 24$, SS; sect. *Sylvestres*) with *N. tomentosiformis* ($2n = 24$, TT; sect. *Tomentosae*) and subsequent chromosome doubling approximately 200,000 years ago (Sheen, 1972; Gray et al., 1974; Kenton et al., 1993; Lim et al., 2000, 2007; Murad et al., 2002, 2004; Chase et al., 2003; Clarkson et al., 2004, 2010). Using these progenitors, it is therefore possible to determine which subgenome of *N. tabacum* is involved in hybrid lethality.

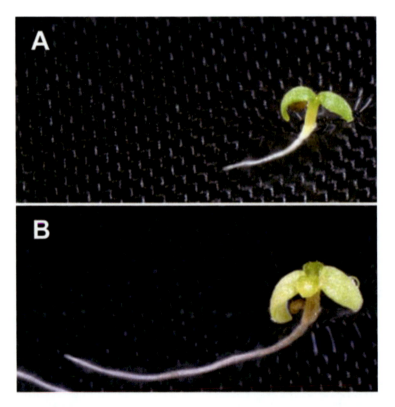

Figure 1. (Continued).

Figure 1. Type II hybrid lethality observed in hybrid seedlings from a cross between *N. suaveolens* and *N. tabacum* 'Samsun NN' at 28°C. (A) Hybrid seeds germinated normally and hybrid seedlings showed apparently normal growth at 3 d after germination (DAG). (B) The hypocotyls and the base of roots turned brown, and the cotyledon color was slightly faded at 7 DAG. (C) The seedlings turned completely brown and died at 15 DAG, although root tips were still white.

In separate studies, *N. suaveolens* (Inoue et al., 1996) and *N. debneyi* (Tezuka et al., 2007) were crossed with the two progenitors of *N. tabacum*. Both *N. suaveolens* and *N. debneyi* produced inviable hybrids showing Type II hybrid lethality in crosses with *N. sylvestris*, whereas the two species produced viable hybrids in crosses with *N. tomentosiformis*. These results clearly indicate that the S subgenome of *N. tabacum* (derived from *N. sylvestris*) is involved in hybrid lethality in *N. suaveolens* × *N. tabacum* and *N. debneyi* × *N. tabacum*.

The particular chromosomes in the *N. tabacum* S subgenome encoding the causal genes were then determined. Each *N. tabacum* chromosome is lettered alphabetically (A–Z, excluding X and Y); chromosomes A–L and M–Z belong to the T and S subgenomes, respectively. A complete set of 24 monosomic lines of *N. tabacum* (Haplo-A–Z), which lack a certain chromosome, has been established in the genetic background of 'Red Russian' (Clausen and Cameron, 1944; Cameron, 1959). These monosomic lines are useful for locating genes on specific chromosomes (Clausen and Cameron, 1944; Gerstel, 1945; Smith, 1968; Kubo, 1982; Kubo et al., 1982; Danehower et al., 1989).

The first application of *N. tabacum* monosomic lines to the study of hybrid lethality in *Nicotiana* involved crosses between *N. tabacum* and *N.*

*africana* (Gerstel et al., 1979). When all 24 monosomic lines were crossed with *N. africana*, only Haplo-H produced a high number of viable hybrids. Based on these results, the H chromosome, which belongs to the T subgenome, was concluded to be related to hybrid lethality. It is not clear, however, whether the viable hybrids from Haplo-H × *N. africana* in that study actually lacked the H chromosome.

Hybrid lethality in crosses between *N. tabacum* and *N. suaveolens* was then analyzed using monosomic *N. tabacum* lines. Ten monosomic lines of the S subgenome (Haplo-M–Z, excluding Haplo-P and Haplo-V) were crossed with *N. suaveolens* (Marubashi and Onosato, 2002). A small number of viable hybrids were obtained only from the cross with Haplo-Q. Because these hybrids possessed 38 or 39 chromosomes, indicating that they lacked the Q chromosome, it was proposed that the Q chromosome is involved in hybrid lethality.

In later studies, Q-chromosome-specific DNA markers (random amplified polymorphic DNA markers, inter-simple sequence repeat markers, and sequence tagged site markers) were developed for *N. tabacum* (Tezuka et al., 2004; Tezuka and Marubashi, 2006b) and used to conclusively prove that the Q chromosome encodes one or more genes causing hybrid lethality in crosses between *N. tabacum* and *N. suaveolens* (Tezuka and Marubashi, 2006b).

Monosomic analyses of hybrid lethality were subsequently conducted in crosses between *N. tabacum* and the 11 other *Suaveolentes* species, namely *N. africana, N. amplexicaulis, N. debneyi, N. excelsior, N. goodspeedii, N. gossei, N. ingulba, N. maritima, N. megalosiphon, N. simulans*, and *N. velutina*. In all cases, the genes causing hybrid lethality were encoded on the Q chromosome of *N. tabacum* (Table 1; Tezuka et al., 2006, 2007, 2010, 2011, 2012). Because *Suaveolentes* species are closely related to each other (Chase et al., 2003; Clarkson et al., 2004, 2010), these results suggest that many species of sect. *Suaveolentes* share the same gene or genes, possibly *Hla1-1* or other alleles of the *Hybrid Lethality A1* (*HLA1*) locus described later in this chapter, triggering hybrid lethality by interaction with genes on the Q chromosome.

Recently, the first linkage map of *N. tabacum* was constructed using simple sequence repeat (SSR) markers (Bindler et al., 2007). The latest version of this linkage map consists of 24 linkage groups corresponding to the haploid chromosome number of *N. tabacum* (Bindler et al., 2011). Linkage group 11 in this map corresponds to the Q chromosome (Tezuka et al., 2012). Based on analyses using a viable Haplo-Q × *N. africana* hybrid seedling with a deletion in a specific Q chromosome region, the gene(s) causing hybrid lethality were localized to a Q chromosome region that is associated with SSR markers PT30342 and PT30365 (Tezuka et al., 2012).

A segregation analysis by classical Mendelian genetics was conducted to identify causal genes in *Suaveolentes* species. *Nicotiana debneyi* and *N. fragrans* were used for the analysis, because *N. debneyi* and *N. fragrans* produce inviable and viable hybrids, respectively, in crosses with *N. tabacum* (Table 1; Yamada et al., 1999; Marubashi and Kobayashi, 2002; Tezuka and Marubashi, 2006a; Tezuka et al., 2007, 2010). $F_1$ hybrids obtained from *N. debneyi* × *N. fragrans* were crossed with *N. tabacum* (Iizuka et al., 2012). Because trispecific hybrids from this cross segregated into inviable and viable hybrids in a 1:1 ratio, it was concluded that *N. debneyi* carries a single dominant gene causing hybrid lethality in the cross with *N. tabacum*. The gene locus was designated as *HLA1*; the *N. debneyi* allele causing hybrid lethality was named *Hla1-1*, and the non-causative allele of *N. fragrans* and *N. tabacum* was tentatively termed *hla1-2* (Iizuka et al., 2012).

The evolution of postzygotic reproductive isolation, such as that due to hybrid sterility and inviability, is often explained by the Bateson–Dobzhansky–Muller model. This model posits that postzygotic isolation is caused by deleterious interaction between at least two genes that have resulted from the accumulation of substitutions in two taxa diverged from a common ancestor (Orr, 1996; Coyne and Orr, 2004). The observed hybrid lethality in crosses between *N. debneyi* and *N. tabacum* is consistent with the Bateson–Dobzhansky–Muller model, because it is caused by epistatic interaction between *HLA1* in *N. debneyi* and one or more genes on the *N. tabacum* Q chromosome.

## Causal Genes of Hybrid Lethality in the Cross between *N. Occidentalis* and *N. Tabacum*

Although the Q chromosome encodes one or more genes responsible for Type II hybrid lethality in most crosses between *Suaveolentes* species and *N. tabacum*, other factors in *N. tabacum* are involved in Type V lethality in crosses with *N. occidentalis*. In crosses of *N. occidentalis* with *N. tabacum* monosomic plants lacking the Q chromosome, hybrid lethality was observed in hybrid seedlings variously lacking or possessing the Q chromosome (Tezuka and Marubashi, 2012). When *N. occidentalis* was crossed with the two progenitors of *N. tabacum*, *N. sylvestris* and *N. tomentosiformis*, hybrid seedlings from *N. occidentalis* × *N. tomentosiformis* and *N. occidentalis* × *N. sylvestris* showed Type V and Type II lethality, respectively (Tezuka and Marubashi, 2012). The type of lethality observed in *N. occidentalis* × *N. tabacum* is the same as that appearing in *N. occidentalis* × *N. tomentosiformis*, strongly suggesting the involvement of one or more genes from the T subgenome.

There is the possibility, however, that S subgenome genes are also responsible for hybrid lethality in *N. occidentalis* × *N. tabacum*, but because lethal symptoms in *N. occidentalis* × *N. tomentosiformis* appear faster than in *N. occidentalis* × *N. sylvestris*, the phenotype of Type V lethality might be preferentially expressed in *N. occidentalis* × *N. tabacum*. If this supposition is true, the hybrid lethality observed in the cross between *N. occidentalis* and *N. tabacum* would be the first reported case of a dual lethal system in the same cross-combination.

Furthermore, the causal chromosome in the *N. tabacum* S subgenome may be chromosome Q: hybrid lethality in *N. occidentalis* × *N. sylvestris* is Type II, which is often observed in crosses between species of sect. *Suaveolentes* and *N. tabacum*. In conclusion, although further evidence is needed, genes from both the S and T subgenomes of *N. tabacum* appear to be responsible for hybrid lethality in the cross between *N. occidentalis* and *N. tabacum*.

## Evolution of Hybrid Lethality in Sect. *Suaveolentes*

Information on phylogeny is critical to estimate the evolutionary order and timing of causal genetic changes underlying reproductive isolation (Moyle and Payseur, 2009). In phylogenetic trees generated by analyses of ITS, plastid, and ncpGS sequence data, species in sect. *Suaveolentes* formed a monophyletic group that was sister to the African species *N. africana* (Chase et al., 2003; Clarkson et al., 2004, 2010).

More recently, Marks et al. (2011) reported the results of phylogenetic analyses of sect. *Suaveolentes* based on a combined data set consisting of morphological characters, chromosome numbers, and coded characters representing strongly-supported nodes in trees generated from individual analyses of ITS, plastid, and ncpGS molecular data. In their tree, which was constructed using *N. africana* as the functional outgroup, the South Pacific species *N. fragrans* was the basal lineage in sect. *Suaveolentes*, with the Australian/South Pacific species *N. debneyi* (New Caledonia, Lord Howe Island, and eastern Australia) in turn sister to all the endemic Australian species.

At least two scenarios can be suggested to explain the evolution of hybrid lethality in sect. *Suaveolentes*. First, progenitors of sect. *Suaveolentes* might have shown Type II hybrid lethality in crosses with *N. tabacum*. Progenitors of sect. *Suaveolentes* have been identified based on ncpGS sequence analysis: the maternal ancestor was *N. sylvestris* and the paternal ancestor was a member of sect. *Trigonophyllae* (Clarkson et al., 2010).

It appears that these progenitors gave rise in South America to the allotetraploid ancestor of sect. *Suaveolentes*, which subsequently dispersed separately to Africa and to Australia. An explosive radiation of taxa occurred only in Australia, which was largely accompanied by dysploid reductions probably due to chromosomal fusions (Clarkson et al., 2004).

Because *N. sylvestris* produces viable hybrids in reciprocal crosses with *N. tabacum* (Christoff, 1928; Kostoff, 1930; East, 1935; Tanaka, 1961), it is probable that either sect. *Trigonophyllae* showed Type II hybrid lethality or the lethality appeared in the allotetraploid ancestor. Additional genetic changes reinforcing postzygotic isolation with *N. tabacum* accumulated in the lineage leading to *N. occidentalis*, giving rise to Type V lethality. In contrast, loss of Type II lethality occurred in lineages leading to *N. benthamiana* and *N. fragrans*.

Another model can be proposed based on the results of recent phylogenetic analyses by Marks et al. (2011). In this model (Figure 2), the allotetraploid ancestor did not cause hybrid lethality after crosses with *N. tabacum*. The appearance of Type II lethality in the lineages leading to *N. africana* and the Australian clade possibly occurred after the divergence of *N. africana*, *N. fragrans*, and the remaining species in sect. *Suaveolentes* (Figure 2).

It is also possible, however, that Type II lethality appeared in the allotetraploid descendent after the divergence of *N. fragrans*, with the allotetraploid descendent subsequently dispersing to Africa and Australia. In Australia, an explosive radiation of taxa occurred that was largely accompanied by dysploid reductions (Clarkson et al., 2004).

During the process, additional genetic changes that reinforced postzygotic isolation with *N. tabacum* accumulated in the lineage leading to *N. occidentalis*, giving rise to Type V lethality (Figure 2). Conversely, loss of Type II lethality gave rise to the lineage leading to *N. benthamiana*.

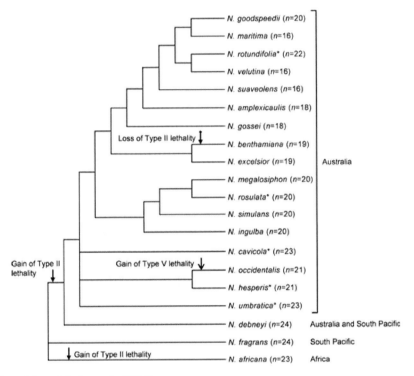

Adapted from Marks et al. (2011).

Figure 2. Schematic phylogenetic tree showing proposed model for gain and loss of hybrid lethality in sect. *Suaveolentes*. The schematic phylogenetic tree including 20 *Suaveolentes* species was drawn based on a tree by Marks et al. (2011). Gains of Type II lethality were established in the lineage leading to *N. africana* and the lineage leading to the Australian clade. Subsequently, gain of Type V lethality occurred in the lineage leading to *N. occidentalis*, and loss of Type II lethality was established in the lineage leading to *N. benthamiana*. Species marked with asterisks are those with known lethality types but with the chromosomes or subgenomes of *N. tabacum* responsible for hybrid lethality undetermined, and thus tentatively included in the schematic phylogenetic tree. Although *N. exigua*, which shows Type II lethality in the cross with *N. tabacum*, was unfortunately not included in the analyses by Marks et al. (2011), this species is likely to reside in the Australian clade based on other phylogenetic studies (Chase et al., 2003; Clarkson et al., 2004, 2010).

# Conclusion

Type II or Type V hybrid lethality is observed in crosses between species in sect. *Suaveolentes* and *N. tabacum*. Previous studies have identified the

*HLA1* locus in *N. debneyi* and determined that the *N. tabacum* Q chromosome and S and T subgenomes encode genes causing hybrid lethality, but none of the causal genes have been cloned and characterized. It is therefore unclear what types of genes control hybrid lethality in *Nicotiana*.

The causal gene for hybrid lethality in *Arabidopsis thaliana* has recently been identified and cloned (Bomblies et al., 2007). In *A. thaliana* intraspecific crosses, hybrid lethality is induced by interaction between an allele of *DANGEROUS MIX 1* (*DM1*) and an allele of *DM2*. *DM1* encodes a *Toll Interleukin Receptor-Nucleotide Binding-Leucine Rich Repeat* (*TIR-NB-LRR*) disease resistance gene homolog. This finding strongly indicates that an autoimmune-like response is the mechanism for hybrid lethality in *A. thaliana* (Bomblies et al., 2007). The causal gene of hybrid lethality in lettuce (*Lactuca saligna* × *L. sativa*) was subsequently also identified as disease-resistance related (Jeuken et al., 2009). Recent findings have thus revealed that at least one type of hybrid lethality is related to disease resistance.

Like Type II and Type V hybrid lethality in crosses between *Nicotiana* sect. *Suaveolentes* and *N. tabacum*, hybrid lethality in *A. thaliana* (Bomblies et al., 2007) and lettuce (Jeuken et al., 2009) is temperature sensitive. Temperature sensitivity is also associated with disease resistance response in *Nicotiana*. The *N* gene, which encodes a TIR-NB-LRR protein (Whitham et al., 1994) and was originally introduced into *N. tabacum* from *N. glutinosa* (Holmes, 1938), confers resistance to tobacco mosaic virus (TMV). This resistance involves a phenomenon known as hypersensitive response (HR). *Nicotiana tabacum* plants possessing *N* show cell death at the site of TMV infection, with TMV restricted to the region surrounding necrotic lesions. This *N*-mediated HR is only induced at temperatures below 28°C (Samuel, 1931; Erickson et al., 1999). Interestingly, when transgenic *N. tabacum* expressing a 50 kDa TMV replicase fragment (*p50*) was crossed with *N. tabacum* homozygous for the *N* gene, the resulting $F_1$ seedlings exhibited systematic necrotic lesions and died at 22°C. At 32°C, $F_1$ seedlings grew normally without necrotic lesions, but they showed necrotic lesions and collapsed after transfer to 22°C (Erickson et al., 1999). I think this resembles hybrid lethality. Considering these findings, disease resistance may be also related to Type II and Type V hybrid lethality. Whether all types of hybrid lethality are explained by disease resistance is uncertain, however. In *Nicotiana*, hybrid lethality Types I–III and V are temperature sensitive, but Type IV is not (Yamada et al., 1999; Tezuka and Marubashi, 2012). To address this question, my colleagues and I are conducting research to elucidate the distribution of the causal genes, and carrying out genetic analyses to identify and clone the causal genes in *Nicotiana*. These studies will reveal the diverse mechanisms of hybrid lethality, contribute to the development of new cultivars, and help in the understanding of speciation mechanisms in *Nicotiana*.

# References

Bai, D.; Reeleder, R. and Brandle, J. E. (1995). Identification of two RAPD markers tightly linked with the *Nicotiana debneyi* gene for resistance to black root rot of tobacco. *Theor. Appl. Genet.*, 91, 1184-1189.

Bindler, G.; van der Hoeven, R.; Gunduz, I.; Plieske, J.; Ganal, M.; Rossi, L.; Gadani, F., and Donini, P. (2007). A microsatellite marker based linkage map of tobacco. *Theor. Appl. Genet.*, 114, 341-349.

Bindler, G.; Plieske, J.; Bakaher, N.; Gunduz, I.; Ivanov, N.; Van der Hoeven, R.; Ganal, M., and Donini, P. (2011). A high density genetic map of tobacco (*Nicotiana tabacum* L.) obtained from large scale microsatellite marker development. *Theor. Appl. Genet.*, 123, 219-230.

Bomblies, K.; Lempe, J.; Epple, P.; Warthmann, N.; Lanz, C.; Dangl, J.L. and Weigel, D. (2007). Autoimmune response as a mechanism for a Dobzhansky-Muller-type incompatibility syndrome in plants. *PLoS Biol.*, 5, e236.

Bomblies, K. and Weigel, D. (2007). Hybrid necrosis: autoimmunity as a potential gene-flow barrier in plant species. *Nat. Rev. Genet.*, 8, 382-393.

Bonnett, H. T.; Kofer, W.; Håkansson, G., and Glimelius, K. (1991). Mitochondrial involvement in petal and stamen development studied by sexual and somatic hybridization of *Nicotiana* species. *Plant Sci.*, 80, 119-130.

Burk, L. G. and Heggestad, H. E. (1966). The genus *Nicotiana*: a source of resistance to diseases of cultivated tobacco. *Econom. Bot.*, 20, 76-88.

Cameron, D. R. (1959). The monosomics of *Nicotiana tabacum*. *Tob. Sci.*, 3, 164-166.

Chase, M. W.; Knapp, S.; Cox, A. V.; Clarkson, J. J.; Butsko, Y.; Joseph, J.; Savolainen, V., and Parokonny, A. S. (2003). Molecular systematics, GISH and the origin of hybrid taxa in *Nicotiana* (Solanaceae). *Ann. Bot.*, 92, 107-127.

Christoff, M. (1928). Cytological studies in the genus *Nicotiana*. *Genetics*, 13, 233-277.

Chung, C. S.; Nakajima, T. and Takeda, G. (1988). Interspecific hybridization between *Nicotiana trigonophylla* Dun. and *N. tabacum* L. through ovule culture. *Jpn. J. Breed.*, 38, 319-326.

Clarkson, J. J.; Knapp, S.; Garcia, V. F.; Olmstead, R. G.; Leitch, A. R., and Chase, M. W. (2004). Phylogenetic relationships in *Nicotiana* (Solanaceae) inferred from multiple plastid DNA regions. *Mol. Phylogenet. Evol.*, 33, 75-90.

Clarkson, J. J.; Kelly, L. J.; Leitch, A. R.; Knapp, S., and Chase, M. W. (2010). Nuclear glutamine synthetase evolution in *Nicotiana*: phylogenetics and the origins of allotetraploid and homoploid (diploid) hybrids. *Mol. Phylogenet. Evol.*, 55, 99-112.

Clausen, R. E. and Cameron, D. R. (1944). Inheritance in *Nicotiana tabacum*. XVIII. Monosomic analysis. *Genetics*, 29, 447-477.

Coyne, J. A. and Orr, H. A. (2004). Speciation. Sinauer Associates, Sunderland, Massachusetts.

Danehower, D. A.; Reed, S. M. and Wernsman, E. A. (1989). Identification of the chromosome carrying the gene for production of β-methylvaleryl sucrose esters in *Nicotiana tabacum*. *Agric. Biol. Chem.*, 53, 2813-2815.

DeVerna, J. W.; Myers, J. R. and Collins, G. B. (1987). Bypassing prefertilization barriers to hybridization in *Nicotiana* using in vitro pollination and fertilization. *Theor. Appl. Genet.*, 73, 665-671.

East, E. M. (1935). Genetic reactions in *Nicotiana*. I. Compatibility. *Genetics*, 20, 403-413.

Erickson, F. L.; Holzberg, S.; Calderon-Urrea, A.; Handley, V.; Axtell, M.; Corr, C., and Baker, B. (1999). The helicase domain of the TMV replicase proteins induces the *N*-mediated defence response in tobacco. *Plant J.*, 18, 67-75.

Fitter, J. T.; Thomas, M. R.; Niu, C., and Rose, R. J. (2005). Investigation of *Nicotiana tabacum* (+) *N. suaveolens* cybrids with carpelloid stamens. *J. Plant Physiol.*, 162, 225-235.

Gerstel, D. U. (1945). Inheritance in *Nicotiana Tabacum*. XIX. Identification of the *Tabacum* chromosome replaced by one from *N. glutinosa* in mosaic-resistant Holmes Samsoun tobacco. *Genetics*, 30, 448-454.

Gerstel, D. U.; Burns, J. A. and Burk, L. G. (1979). Interspecific hybridizations with an African tobacco, *Nicotiana africana* Merxm. *J. Hered.*, 70, 342-344.

Goodspeed, T. H. (1954). The genus *Nicotiana*. Chronica Botanica Company, Waltham, Massachusetts.

Gray, J. C.; Kung, S. D.; Wildman, S. G., and Sheen, S. J. (1974). Origin of *Nicotiana tabacum* L. detected by polypeptide composition of Fraction I protein. *Nature*, 252, 226-227.

Holmes, F. O. (1938). Inheritance of resistance to tobacco-mosaic disease in tobacco. *Phytopathology*, 28, 553-561.

Iizuka, T.; Oda, M. and Tezuka, T. (2010). Hybrid lethality expressed in hybrids between *Nicotiana tabacum* and 50 lines of 21 wild species in section *Suaveolentes*. *Breed. Res.*, 12 (Suppl. 1), 244 (in Japanese).

Iizuka, T.; Kuboyama, T.; Marubashi, W.; Oda, M., and Tezuka, T. (2012). *Nicotiana debneyi* has a single dominant gene causing hybrid lethality in crosses with *N. tabacum*. *Euphytica*, 186, 321-328.

Ilcheva, V.; San, L. H.; Zagorska, N., and Dimitrov, B. (2001). Production of male sterile interspecific somatic hybrids between transgenic *N. tabacum* (*bar*) and *N. rotundifolia* (*npt II*) and their identification by AFLP analysis. *In Vitro Cell. Dev. Biol. Plant*, 37, 496-502.

Inoue, E.; Marubashi, W. and Niwa, M. (1994). Simple method for overcoming the lethality observed in the hybrid between *Nicotiana suaveolens* and *N. tabacum*. *Breed. Sci.*, 44, 333-336.

Inoue, E.; Marubashi, W. and Niwa, M. (1996). Genomic factors controlling the lethality exhibited in the hybrid between *Nicotiana suaveolens* Lehm. and *N. tabacum* L. *Theor. Appl. Genet.*, 93, 341-347.

Inoue, E.; Marubashi, W. and Niwa, M. (1997). Improvement of the method for overcoming the hybrid lethality between *Nicotiana suaveolens* and *N. tabacum* by culture of $F_1$ seeds in liquid media containing cytokinins. *Breed. Sci.*, 47, 211-216.

Iwai, S.; Kishi, C.; Nakata, K., and Kubo, S. (1985). Production of a hybrid of *Nicotiana repanda* Willd. × *N. tabacum* L. by ovule culture. *Plant Sci.*, 41, 175-178.

Iwai, S.; Kishi, C.; Nakata, K., and Kawashima, N. (1986). Production of *Nicotiana tabacum* × *Nicotiana acuminata* hybrid by ovule culture. *Plant Cell Rep.*, 5, 403-404.

Jeuken, M. J. W.; Zhang, N. W.; McHale, L. K.; Pelgrom, K.; den Boer, E.; Lindhout, P.; Michelmore, R. W.; Visser, R. G. F., and Niks, R. E. (2009). *Rin4* causes hybrid necrosis and race-specific resistance in an interspecific lettuce hybrid. *Plant Cell*, 21, 3368-3378.

Kenton, A.; Parokonny, A. S.; Gleba, Y. Y., and Bennett, M. D. (1993). Characterization of the *Nicotiana tabacum* L. genome by molecular cytogenetics. *Mol. Gen. Genet.*, 240, 159-169.

Kincaid, R. R. (1949). Three interspecific hybrids of tobacco. *Phytopathology*, 39, 284-287.

Kitamura, S.; Inoue, M.; Ohmido, N.; Fukui, K., and Tanaka, A. (2003). Chromosomal rearrangements in interspecific hybrids between *Nicotiana gossei* Domin and *N. tabacum* L., obtained by crossing with pollen exposed to helium ion beams or gamma-rays. *Nucl. Instrum. Methods Phys. Res. B*, 206, 548-552.

Knapp, S.; Chase, M. W. and Clarkson, J. J. (2004). Nomenclatural changes and a new sectional classification in *Nicotiana* (Solanaceae). *Taxon*, 53, 73-82.

Kostoff, D. (1930). Ontogeny, genetics, and cytology of *Nicotiana* hybrids. *Genetica*, 12, 33-139.

Kubo, T. (1982). Identification of the chromosome carrying the gene for virescence in tobacco. *Bulletin of the Iwata Tobacco Experiment Station*, 14, 23-27 (in Japanese with English summary).

Kubo, T.; Sato, M.; Tomita, H., and Kawashima, N. (1982). Identification of the chromosome carrying the gene for *cis*-abienol production by the use of monosomics in *Nicotiana tabacum* L. *Tob. Sci.*, 26, 126-128.

Kubo, T.; Kumashiro, T. and Saito, Y. (1988). Cytoplasmic male sterile lines of a tobacco variety, Tsukuba 1, developed by asymmetric protoplast fusion. *Jpn. J. Breed.*, 38, 158-164.

Kuboyama, T.; Chung, C. S. and Takeda, G. (1994). The diversity of interspecific pollen-pistil incongruity in *Nicotiana*. *Sex. Plant Reprod.*, 7, 250-258.

Kuboyama, T. and Takeda, G. (2000). Genomic factors responsible for abnormal morphology of pollen tubes in the interspecific cross *Nicotiana tabacum* × *N. rustica*. *Sex. Plant Reprod.*, 12, 333-337.

Kumashiro, T. and Kubo, T. (1986a). Cytoplasm transfer of *Nicotiana debneyi* to *N. tabacum* by protoplast fusion. *Jpn. J. Breed.*, 36, 39-48.

Kumashiro, T. and Kubo, T. (1986b). Stability of agronomic traits in cytoplasmic male sterile tobacco obtained by protoplast fusion. *Jpn. J. Breed.*, 36, 284-290.

Kumashiro, T.; Asahi, T. and Komari, T. (1988). A new source of cytoplasmic male sterile tobacco obtained by fusion between *Nicotiana tabacum* and X-irradiated *N. africana* protoplasts. *Plant Sci.*, 55, 247-254.

Ladiges, P. Y.; Marks, C. E. and Nelson, G. (2011). Biogeography of *Nicotiana* section *Suaveolentes* (Solanaceae) reveals geographical tracks in arid Australia. *J. Biogeogr.*, 38, 2066-2077.

Laskowska, D. and Berbeć, A. (2012). Production and characterization of amphihaploid hybrids between *Nicotiana wuttkei* Clarkson et Symon and *N. tabacum* L. *Euphytica*, 183, 75-82.

Leitch, I. J.; Hanson, L.; Lim, K. Y.; Kovarik, A.; Chase, M. W.; Clarkson, J. J., and Leitch, A. R. (2008). The ups and downs of genome size evolution in polyploid species of *Nicotiana* (Solanaceae). *Ann. Bot.*, 101, 805-814.

Li, B. C.; Bass, W. T. and Cornelius, P. L. (2006). Resistance to tobacco black shank in *Nicotiana* species. *Crop Sci.*, 46, 554-560.

Lim, K. Y.; Matyášek, R.; Lichtenstein, C. P., and Leitch, A. R. (2000). Molecular cytogenetic analyses and phylogenetic studies in the *Nicotiana* section Tomentosae. *Chromosoma*, 109, 245-258.

Lim, K. Y.; Kovarik, A.; Matyasek, R.; Chase, M. W.; Clarkson, J. J.; Grandbastien, M. A., and Leitch, A. R. (2007). Sequence of events leading to near-complete genome turnover in allopolyploid *Nicotiana* within five million years. *New Phytol.*, 175, 756-763.

Lloyd, R. (1975). Tissue culture as a means of circumventing lethality in an interspecific *Nicotiana* hybrid. *Tob. Sci.*, 19, 4-6.

Manabe, T.; Marubashi, W. and Onozawa, Y. (1989). Temperature-dependent conditional lethality in interspecific hybrids between *Nicotiana suaveolens* Lehm. and *N. tabacum* L. *Proceedings of the 6th International Congress of SABRAO*, pp. 459-462, Tsukuba, Japan, August 1989.

Marks, C. E.; Newbigin, E. and Ladiges, P. Y. (2011). Comparative morphology and phylogeny of *Nicotiana* section Suaveolentes (Solanaceae) in Australia and the South Pacific. *Aust. Syst. Bot.*, 24, 61-86.

Marubashi, W. and Nakajima, T. (1985). Overcoming cross-incompatibility between *Nicotiana tabacum* L. and *N. rustica* L. by test-tube pollination and ovule culture. *Jpn. J. Breed.*, 35, 429-437.

Marubashi, W. and Kobayashi, M. (2002). Apoptosis detected in hybrids between *Nicotiana debneyi* and *N. tabacum*. *Breed. Res.*, 4, 209-214 (in Japanese with English summary).

Marubashi, W. and Onosato, K. (2002). Q chromosome controls the lethality of interspecific hybrids between *Nicotiana tabacum* and *N. suaveolens*. *Breed. Sci.*, 52, 137-142.

McCray, F. A. (1932). Compatibility of certain *Nicotiana* species. *Genetics*, 17, 621-636.

McCray, F. A. (1933). Embryo development in *Nicotiana* species hybrids. *Genetics*, 18, 95-110.

Merxmüller, H. and Buttler, K. P. (1975). Nicotiana in der afrikanischen Namib-ein pflanzengeographisches und phylogenetisches Rätsel. *Mitt. Bot. Staatssamml. München*, 12, 91-104.

Mino, M.; Maekawa, K.; Ogawa, K.; Yamagishi, H., and Inoue, M. (2002). Cell death process during expression of hybrid lethality in interspecific $F_1$ hybrid between *Nicotiana gossei* Domin and *Nicotiana tabacum*. *Plant Physiol.*, 130, 1776-1787.

Moyle, L. C. and Payseur, B. A. (2009). Reproductive isolation grows on trees. *Trends Ecol. Evol.*, 24, 591-598.

Murad, L.; Lim, K. Y.; Christopodulou, V.; Matyasek, R.; Lichtenstein, C. P.; Kovarik, A., and Leitch, A. R. (2002). The origin of tobacco's T genome is traced to a particular lineage within *Nicotiana tomentosiformis* (Solanaceae). *Am. J. Bot.*, 89, 921-928.

Murad, L.; Bielawski, J. P.; Matyasek, R.; Kovarík, A.; Nichols, R. A.; Leitch, A. R., and Lichtenstein, C. P. (2004). The origin and evolution of geminivirus-related DNA sequences in *Nicotiana*. *Heredity*, 92, 352-358.

Nagao, T. (1978). Somatic hybridization by fusion of protoplasts. I. The combination of *Nicotiana tabacum* and *Nicotiana rustica*. *Jpn. J. Crop Sci.*, 47, 491-498 (in Japanese with English summary).

Nagao, T. (1979). Somatic hybridization by fusion of protoplasts. II. The combinations of *Nicotiana tabacum* and *N. glutinosa* and of *N. tabacum* and *N. alata*. *Jpn. J. Crop Sci.*, 48, 385-392 (in Japanese with English summary).

Nagao, T. (1982). Somatic hybridization by fusion of protoplasts. III. Somatic hybrids of sexually incompatible combinations *Nicotiana tabacum* + *Nicotiana repanda* and *Nicotiana tabacum* + *Salpiglossis sinuata*. *Jpn. J. Crop Sci.*, 51, 35-42.

Nikova, V. M. and Zagorska, N. A. (1990). Overcoming hybrid incompatibility between *Nicotiana africana* Merxm. and *N. tabacum* and development of cytoplasmically male sterile tobacco forms. *Plant Cell Tiss. Org. Cult.*, 23, 71-75.

Nikova, V. M.; Zagorska, N. A. and Pundeva, R. S. (1991). Development of four tobacco cytoplasmic male sterile sources using *in vitro* techniques. *Plant Cell Tiss. Org. Cult.*, 27, 289-295.

Nikova, V.; Pundeva, R. and Petkova, A. (1999). *Nicotiana tabacum* L. as a source of cytoplasmic male sterility in interspecific cross with *N. alata* Link and Otto. *Euphytica*, 107, 9-12.

Orr, H. A. (1996). Dobzhansky, Bateson, and the genetics of speciation. *Genetics*, 144, 1331-1335.

Pandey, K. K. (1977). Mentor pollen: possible role of wall-held pollen growth promoting substances in overcoming intra- and interspecific incompatibility. *Genetica*, 47, 219-229.

Pittarelli, G. W. and Stavely, J. R. (1975). Direct hybridization of *Nicotiana repanda* × *N. tabacum*. *J. Hered.*, 66, 281-284.

Reed, S. M. and Collins, G. B. (1978). Interspecific hybrids in *Nicotiana* through in vitro culture of fertilized ovules. *J. Hered.*, 69, 311-315.

Reed, S. M. and Collins, G. B. (1980). Histological evaluation of seed failure in three *Nicotiana* interspecific hybrids. *Tob. Sci.*, 24, 154-156.

Rieseberg, L. H. and Blackman, B. K. (2010). Speciation genes in plants. *Ann. Bot.*, 106, 439-455.

Samuel, G. (1931). Some experiments on inoculating methods with plant viruses, and on local lesions. *Ann. Appl. Biol.*, 18, 494-507.

Sheen, S. J. (1972). Isozymic evidence bearing on the origin of *Nicotiana tabacum* L. *Evolution*, 26, 143-154.

Shintaku, Y.; Yamamoto, K. and Nakajima, T. (1988). Interspecific hybridization between *Nicotiana repanda* Willd. and *N. tabacum* L. through the pollen irradiation technique and the egg cell irradiation technique. *Theor. Appl. Genet.*, 76, 293-298.

Shintaku, Y.; Yamamoto, K. and Takeda, G. (1989). Chromosomal variation in hybrids between *Nicotiana repanda* Willd. and *N. tabacum* L. through pollen and egg-cell irradiation techniques. *Genome*, 32, 251-256.

Shizukuda, N. and Nakajima, T. (1982). Productioin of interspecific hybrids between *Nicotiana rustica* L. and *N. tabacum* L. through ovule culture. *Jpn. J. Breed.*, 32, 371-377 (in Japanese with English summary).

Smith, H. H. (1968). Recent cytogenetic studies in the genus *Nicotiana*. In: *Advances in Genetics Vol. 14*. Academic Press, New York, pp. 1-54.

Stavely, J. R.; Pittarelli, G. W. and Burk, L. G. (1973). *Nicotiana repanda* as a potential source for disease resistance in *N. tabacum*. *J. Hered.*, 64, 265-271.

Stebbins, G. L. (1966). Reproductive isolation and the origin of species. In: *Processes of organic evolution*. Prentice-Hall, New Jersey, pp. 85-112.

Subhashini, U.; Venkateswarlu, T.; Anjani, K., and Prasad, G. S. R. (1985). In vitro hybridization in an incompatible cross *Nicotiana glutinosa* × *Nicotiana megalosiphon*. *Theor. Appl. Genet.*, 71, 545-549.

Subhashini, U.; Venkateswarlu, T. and Anjani, K. (1986). Embryo rescue in *Nicotiana* hybrids by in-vitro culture. *Plant Sci.*, 43, 219-222.

Sun, Y. H.; Xue, Q. Z.; Ding, C. M.; Zhang, X. Y.; Zhang, L. L.; Wang, W. F. and Ali, S. (2005). Somatic cybridization between *Nicotiana tabacum* and

*N. repanda* based on a single inactivation procedure of nuclear donor parental protoplasts. *Plant Sci.*, 168, 303-308.

Swaminathan, M. S. (1957). One-way incompatibility in some species crosses in the genus *Nicotiana*. *Indian J. Genet. Plant Breed.*, 17, 23-26.

Swaminathan, M. S. and Murty, B. R. (1959). Effect of X-radiation on pollen tube growth and seed setting in crosses between *Nicotiana tabacum* and *N. rustica*. *Z. Vererbungsl.*, 90, 393-399.

Tanaka, M. (1961). The effect of irradiated pollen grains on species crosses of *Nicotiana*. *Bulletin of the Hatano Tobacco Experiment Station*, 51, 1-38 (in Japanese with English summary).

Ternovskii, M. F.; Butenko, R. G. and Moiseeva, M. E. (1972). The use of tissue culture to overcome the barrier of incompatibility between species and sterility of interspecies hybrids. *Sov. Genet.*, 8, 27-33. Translated from *Genetika*, 8, 38-45.

Ternovskii, M. F.; Shinkareva, I. K. and Lar'kina, N. I. (1976). Production of interspecific tobacco hybrids by the pollination of ovules in vitro. *Sov. Genet.*, 12, 1209-1213. Translated from *Genetika*, 12, 40-45.

Tezuka, T. and Marubashi, W. (2004). Apoptotic cell death observed during the expression of hybrid lethality in interspecific hybrids between *Nicotiana tabacum* and *N. suaveolens*. *Breed. Sci.*, 54, 59-66.

Tezuka, T.; Onosato, K.; Hijishita, S., and Marubashi, W. (2004). Development of Q-chromosome-specific DNA markers in tobacco and their use for identification of a tobacco monosomic line. *Plant Cell Physiol.*, 45, 1863-1869.

Tezuka, T. and Marubashi, W. (2006a). Genomic factors lead to programmed cell death during hybrid lethality in interspecific hybrids between *Nicotiana tabacum* and *N. debneyi*. *SABRAO J. Breed. Genet.*, 38, 69-81.

Tezuka, T. and Marubashi, W. (2006b). Hybrid lethality in interspecific hybrids between *Nicotiana tabacum* and *N. suaveolens*: evidence that the Q chromosome causes hybrid lethality based on Q-chromosome-specific DNA markers. *Theor. Appl. Genet.*, 112, 1172-1178.

Tezuka, T.; Kuboyama, T.; Matsuda, T., and Marubashi, W. (2006). Expression of hybrid lethality in interspecific crosses between *Nicotiana tabacum* and nine wild species of section *Suaveolentes*, and the chromosome responsible for hybrid lethality. *Breed. Res.*, 8 (Suppl. 2), 139 (in Japanese).

Tezuka, T.; Kuboyama, T.; Matsuda, T., and Marubashi, W. (2007). Possible involvement of genes on the Q chromosome of *Nicotiana tabacum* in expression of hybrid lethality and programmed cell death during interspecific hybridization to *Nicotiana debneyi*. *Planta*, 226, 753-764.

Tezuka, T.; Kuboyama, T.; Matsuda, T., and Marubashi, W. (2010). Seven of eight species in *Nicotiana* section *Suaveolentes* have common factors leading to hybrid lethality in crosses with *Nicotiana tabacum*. *Ann. Bot.*, 106, 267-276.

Tezuka, T.; Matsuo, C.; Okamori, T.; Iizuka, T.; Oda, M., and Marubashi, W. (2011). The Q chromosome is responsible for hybrid lethality in crosses between *Nicotiana tabacum* and *Nicotiana simulans*. *Abstracts of the $8^{th}$ Solanaceae and $2^{nd}$ Cucurbitaceae Genome Joint Conference*, pp. 90, Kobe, Japan, November-December 2011.

Tezuka, T. (2012). Hybrid lethality in the genus *Nicotiana*. In: Mworia, J. K. (Ed), *Botany*. InTech, Rijeka, Croatia, pp. 191-210. Available from: http://www.intechopen.com/articles/show/title/hybrid-lethality-in-the-genus-nicotiana

Tezuka, T. and Marubashi, W. (2012). Genes in S and T subgenomes are responsible for hybrid lethality in interspecific hybrids between *Nicotiana tabacum* and *Nicotiana occidentalis*. *PLoS ONE*, 7, e36204.

Tezuka, T.; Matsuo, C.; Iizuka, T.; Oda, M. and Marubashi, W. (2012). Identification of *Nicotiana tabacum* linkage group corresponding to the Q chromosome gene(s) involved in hybrid lethality. *PLoS ONE*, 7, e37822.

Trojak-Goluch, A. and Berbeć, A. (2007). Meiosis and fertility in interspecific hybrids of *Nicotiana tabacum* L. × *N. glauca* Grah. and their derivatives. *Plant Breed.*, 126, 201-206.

Whitham, S.; Dinesh-Kumar, S. P.; Choi, D.; Hehl, R.; Corr, C., and Baker, B. (1994). The product of the tobacco mosaic virus resistance gene *N*: similarity to toll and the interleukin-1 receptor. *Cell*, 78, 1101-1115.

Yamada, T.; Marubashi, W. and Niwa, M. (1999). Detection of four lethality types in interspecific crosses among *Nicotiana* species through the use of three rescue methods for lethality. *Breed. Sci.*, 49, 203-210.

Zhou, W. M.; Yoshida, K.; Shintaku, Y., and Takeda, G. (1991). The use of IAA to overcome interspecific hybrid inviability in reciprocal crosses between *Nicotiana tabacum* L. and *N. repanda* Willd. *Theor. Appl. Genet.*, 82, 657-661.

In: Herbaceous Plants
Editor: Florian Wallner

ISBN: 978-1-62618-729-0
© 2013 Nova Science Publishers, Inc.

*Chapter 4*

# Tolerance of Herbaceous Plants to Multiple Contaminations in an Industrial Barren Near a Nickel-Copper Smelter

*R. Kikuchi[1], T. T. Gorbacheva[2], M. V. Slukovskaya[3] and L. A. Ivanova[4]*

[1]CERNAS – Instituto Politécnico de Coimbra/
Ryukoku University, Portugal
[2]Institute of the North Industrial Ecological Problem, Kola Science Center,
Russian Academy of Sciences, Apatity, Russia
[3]Petrozavodsk State University, Republid of Karelia, Russia
[4]Polar-Alpine Botanical Garden-Institute,
Kirovsk-Kola Science Center, Kirovsk, Russia

## Abstract

A better understanding of heavy metal sources, their accumulation in the soil and the effect of their presence in soil on plant systems seems to be particularly important in present-day research on risk assessments. It is necessary to evaluate plant tolerance when discussing this subject.

The most severe effects of metals on ecosystems are from local pollution in the Arctic/subarctic regions, and the Kola Peninsula

(66–70°N and 28°30'–41°30'E) in Russia is one of the most seriously polluted regions: close by nickel-copper smelters, the deposition of metal pollutants has severely damaged the soil and ground vegetation, resulting in an industrial barren. During 2011–2012, a field test was performed near the smelter complex (67°51'N, 32°48'E). The applied method is based on cultivation of perennial grasses using hydroponics with vermiculite from a local deposit followed by rolled lawn placement on the metal-polluted sites. To avoid root system disturbance, we used an additional 5 cm-layer from local deposit carbonatites (lime-like materials). Original carbonatites show high initial nutritional status: bioavailable forms of 18 mg kg$^{-1}$ K, 123 g kg$^{-1}$ Ca, 1.8 g kg$^{-1}$ Mg and 89 mg kg$^{-1}$ P. Multiple pollution was observed during the field test: the precipitation amount of $SO_4^{2-}$ in the study field was over 5.57 times (4406,3 g ha$^{-1}$) greater than that in the background field, the Cu amount in the study field was over 645 times (572,5 g ha$^{-1}$) greater than that in the background field, and the Ni amount in the study field was over 824 times (685,3 g ha$^{-1}$) greater than that in the background field. The results obtained from leaf diagnostics also show that the monitored plants are tolerant to multiple stress (cf. monitored zone vs. background zone): N – 15942 mg kg$^{-1}$ vs. 11300 mg kg$^{-1}$; P – 608 mg kg$^{-1}$ vs. 1660 mg kg$^{-1}$; K – 17266 mg kg$^{-1}$ vs.12290 mg kg$^{-1}$; Ca – 5388 mg kg$^{-1}$ vs. 1700 mg kg$^{-1}$; Mg – 1947 mg kg$^{-1}$ vs. 650 mg kg$^{-1}$. The authors' observation is still continuing in order to study the influence of freezing and the nutrient loss rate.

# Introduction

The problem of environmental pollution due to toxic metals has begun to cause concern in many industrial zones, and toxic heavy metals entering the ecosystem may lead to geoaccumulation, bioaccumulation and biomagnification (Ward, 1995). Heavy metals like Cu, Zn, Ni and other trace elements are important for proper functioning of biological systems and their deficiency or excess could lead to a number of disorders (Ward, 1995). Food chain contamination by heavy metals has become a burning issue in recent years because of their potential accumulation in biosystems through contaminated water, soil and air. Therefore, a better understanding of heavy metal sources, their accumulation in the soil and the effect of their presence in water and soil on plant systems seem to be particularly important issues of present-day research on risk assessments [review in (Lokeshwari and Chandrappa, 2006)].

Retarded growth is commonly considered as one of the very general and best known plant responses to industrial emissions including sulfur dioxide, fluorine and heavy metals (Treshow and Anderson, 1989), although "positive" effects of these pollutants have also been documented (Lechowicz, 1987). Textbooks and review papers do not discriminate between growth responses of woody and herbaceous plants but consider all plants together (e.g. (Bell and Treshow, 2002; Lechowicz, 1987)). For obvious reasons, manipulative studies more frequently employ short-living herbaceous plants than long-living woody species as test organisms. These studies – conducted both in fully controlled environments and under field conditions – report adverse effects of different pollutants on the growth of herbaceous plants (Hassan, 2004; Ryser and Sauder, 2006). In contrast to experimental studies, reports on the responses of field-growing herbaceous plants to the local severe impact of industrial emissions are surprisingly scanty (Kozlov and Zvereva, 2007a). Therefore, it is necessary to evaluate plant tolerance in a field facing severe pollution. This chapter reports a field survey of herbaceous plants suffering from multiple metal contaminations. Basic information is briefly reviewed first, followed by a description of the field survey.

# Effects of Metal Pollution

A consequence of heavy metal pollution of nature is the contamination of plants by heavy metals (Wu et al., 1998); this contamination may threaten the health of animals and human beings via the food chain (Wang et al., 2001). Many researchers have examined the inhibitory effect of heavy metal compounds on plant growth, and they have reported that there are two aspects of the interaction of plants with heavy metals: (i) heavy metals show negative effects on plants, and (ii) plants have their own resistance mechanisms against toxic effects and for detoxifying heavy metal pollution (review in (Ping, 2003)).

## Negative Effects

Both natural and anthropogenic processes and sources emit metals into air and water. Plants and animals depend on some metals as micronutrients. However; certain forms of some metals can also be toxic, even in relatively small amounts, and therefore pose a risk to the environment (Kikuchi and

Gorbacheva, 2006). The effects of heavy metals on plant growth are outlined below on the basis of published data (Ping, 2003).

Heavy metals such as Cd and Pb are non-essential elements for plants. If plentiful amounts are accumulated in plants, heavy metals will adversely affect the absorption and transport of essential elements, disturb metabolism, and have an impact on growth and reproduction. The germinating ratio and growth rate of barley decline, for instance, when polluted with Cd, and the decline is related to the dosage and duration. Bean seedlings become brown and die under Cd stress. The roots are one of the target organs of Cd pollution, so that the root growth of crops such as wheat, maize, pumpkin, cucumber and garlic *(Allium sativum* L) are inhibited.

## Tolerance

Heavy metals have an impact on the growth of plants. However, plants also have mechanisms to detoxify heavy metals. These mechanisms are reviewed with regard to their individual aspects (Ping, 2003): (i) retard the absorption of heavy metals and reduce the content of heavy metals in plants and, therefore, abate the impact of heavy metals on the plant; (ii) store the absorbed heavy metals in trichomes of epidermis to avoid the direct effect of heavy metals on the mesophyll; (iii) precipitate and chelate heavy metals in a special site in the plant for detoxification; (iv) increase anti-oxidization enzyme activities and remove free radicals to prevent damage to the plant by free radicals; generally, under heavy metal stress, many free radicals are produced; and (v) produce many kinds of protein, which are induced by heavy metals, to resist the impact of environmental stress on the plant (Zhang et al., 1999).

An example of plant resistance is given: the effects of Cd and Pb in soil on rice and cotton indicate that the ability of genetically modified cotton to resist metal damage is stronger than that found for common cotton under the same conditions (Qin et al. 2000).

## Bioavailability

The bioavailability and effects of heavy metals on plants are influenced by many factors, such as organic substances, fertilization, plant species and so on (review in (Ping, 2003)): (i) organic and amino acids (such as citric acid, tartaric acid, oxalic acid, succinic acid, aspartic acid and glutamic acid) excreted by the plant roots form soluble complexes with heavy metals and these acids increase; (ii) pH impacts the behavior of heavy metals, so acid rain might reduce the content of heavy metals in soil; (iii) effects of heavy metals on the growth, and physiological and biochemical characteristics of plants heighten the effects of other metal elements and influence the accumulation of heavy metals by plants; and (iv) some elements affect the absorption of heavy metals in plants e.g. the interaction of Zn and Cd on absorption by rape seed seedlings (*Brassica rape*), and it increases the content of heavy metals in leachate.

# Field Survey

As stated in the introduction, researches on the responses of field-growing herbaceous plants to local severe impact of industrial emissions are scanty; it is therefore essential to plan a field survey in a region which is currently suffering from metal pollution. Close by the nickel-copper smelters situated in the Arctic region, the deposition of metal pollutants has severely damaged the soil and ground vegetation (AMAP Assessment, 2006).

The polluted land near the "Severonikel" smelter in the Monchegorsk area (67°51'N, 32°48'E) on the Kola Peninsula (66-70°N and 28°30'-41°30'E) in Russia was proposed as the location for a field survey (Figure 1). The experimental site (67°55.783'N, 32°51.535'E) was 0.7 km from a source of emissions.

Native vegetation was absent from the site, and the ground resembled modified peat with high contents of Cu and Ni in forms available to plants, sufficiently exceeding maximum permissible concentrations in the Russian Federation (on the order of 2 and 40 times respectively). The pH value of soil water was 3.6-4.0 and the peat humidity was 40-60%.

Since the authors' research project started in 2011 and will continue in 2013, the final results have not yet been obtained. Some interim results obtained during 2011-2012 are presented here.

The nickel-copper smelter at Monchegorsk is one of the largest European sources of aerial pollution, predominantly of sulfhur dioxide and heavy metals (Table 1), and the only big polluter in the central part of the Kola Peninsula. The nickel-copper smelter at Nikel and the ore-roasting factory in Zapolyarny emit similar amounts of pollutants (Table 1). As is seen in Table 1, the study field is suffering from multiple pollution elements such as $SO_2$, Ni and Cu.

Figure 1. Location of study field and whole of study field at the initial stage.

The long-lasting and severe impact of these three polluting elements resulted in contamination of over 15,000 km$^2$, forest decline on some 400-600 km$^2$ and formation of extensive (20-30 km$^2$) industrial barrens around each of these emission sources (Kozlov and Zvereva, 2007b). The zones of invisible plant damage extend up to 80-100 km (Tikkanen and Niemelä, 1995). Vegetation at the most polluted sites was generally similar near all these smelters (bare ground with occasional dwarf shrub patches) (Rigina and Kozlov, 2000).

Table 1. Emissions (ton/year) of chief pollutants on the Kola Peninsula (Berlyand, 1991; Milyaev and Yasenskij, 2004) (more recent data are not yet available)

| Polluter | SO$_2$ | | Ni | | Cu | |
|---|---|---|---|---|---|---|
| | 1990 | 2003 | 1990 | 2003 | 1990 | 2003 |
| Nikel | 190,100 | 60,600 | 140 | 150 | 90 | 80 |
| Zapolyarny | 67,300 | 63,700 | 160 | 180 | 90 | 80 |
| Monchegorsk | 232,600 | 42,100 | 2,710 | 910 | 1,810 | 700 |

## Material and Method

Long-term rehabilitation experiments shows that minimal intervention promotes natural succession which can be used instead of re-greening or an artificial re-creation of the desired ecosystem (Kozlov and Zvereva, 2007b), and it is considered that the proper application of soil additive such as vermiculite and carbonatites will contribute to creating grass cover of tolerant vegetation.

Therefore, a field test using rolled lawn and direct sowing was planned in order to study how to properly apply both methods of artificial grass cover to restore the degraded forest land affected by severe pollution (cf. Table 1) and climatic conditions (annual mean temperature of -1.0°C).

## Preparation of Study Field

The selected field (cf. Figure 1) for this study was prepared according to the scheme summarized in Table 2, and the field was divided originally into 26 single plots of 1.0 m$^2$ with about 0.5 m distance from one to another (cf. Figure 2).

Figure 2. Field prepared according to the test scheme (cf. Table 2) in June 2011.

**Table 2. Preparation scheme of plantation in the study field**

| Test No. | Ground | Permeable barrier | Creation of grass cover | Plot quantity |
|---|---|---|---|---|
| 1 (control) | Peat | Not applied | direct seeding | 2 |
|  |  |  | rolled lawn | 2 |
| 2 (control) |  | Sand from local sand-pit | direct seeding | 3 |
|  |  |  | rolled lawn | 3 |
| 3 |  | Carbonatite | direct seeding | 4 |
|  |  |  | rolled lawn | 4 |
| 4 | Peat in depression |  | direct seeding | 1 |
|  |  |  | rolled lawn | 1 |
| 5 | Mineral layer |  | direct seeding | 3 |
|  |  |  | rolled lawn | 3 |

A polyethylene film (~1.2 m width) was put down on the ground, and both edges of this film were twisted up to form a ditch (~1.0 m width) in the center part of the film. The formed ditch was filled with vermiculite (~1 mm size, 98% porosity, ~2.5 g cm$^{-3}$ density and ~500 kg m$^{-3}$ filling mass) to make a substratum of 1 cm height. After prepared liquid fertilizer was spread on the substratum surface, the fertilized substratum was allowed to settle for 30 minutes. The applied fertilizer contained the following elements: 141.6 mg $\ell^{-1}$ N, 37.8 mg $\ell^{-1}$ P, 193 mg $\ell^{-1}$ K, 164.5 mg $\ell^{-1}$ Ca, 29.7 mg $\ell^{-1}$ Mg, 97.8 mg $\ell^{-1}$ S, 4.0 mg $\ell^{-1}$ Fe, 0.5 mg $\ell^{-1}$ B, 0.05 mg $\ell^{-1}$ Zn, 0.01 mg $\ell^{-1}$ Mo and 0.01 mg $\ell^{-1}$ Co. After such saturation, the vermiculite substratum had a high initial nutritional status: the contents in bioavailable form were 471 mg kg$^{-1}$ K, 2.6 g kg$^{-1}$ Ca, 10.8 g kg$^{-1}$ Mg and 480 mg kg$^{-1}$ P.

Experiments without any barrier and with sand from a local sand-pit were conducted as a control. It should be mentioned that sand from the local sand-pit had some nutritional status: the contents in bioavailable form were 254 mg kg$^{-1}$ K, 2.5 g kg$^{-1}$ Ca, 170 mg kg$^{-1}$ Mg and 38 mg kg$^{-1}$ P.

*Grass Preparation*

The cultivated grass seeds consisted of *Agropyron intermedium* (Host.) Beauv., *Festuca rubra* L., *Lolium perenne* L. and *Phleum pratense* L. in mass ratio of 1:1:2:2. The seeding dose per m$^2$ was the same in both the rolled lawn and direct sowing. Such seeding mix is considered to be better able to survive in severe conditions of technogenic load and under the unfavorable climatic conditions of the Polar region (Ivanova et al., 2010).

The vermiculite substratum prepared according to the hydroponics method described above was covered with film to promote grass seed mix germination in greenhouse conditions (Figure 3a). The cultivated grass seeds sprouted from 22 June 2011 during a period of two weeks (Figure 3b). Each grass dose in rolled lawn was 19 g m$^{-2}$ *Agropyron intermedium* (Host.) Beauv., 19 g m$^{-2}$ *Festuca rubra* L., 38 g m$^{-2}$ *Lolium perenne* L. and 38 g m$^{-2}$. *Phleum pratense* L. The total density of rolled lawn after seed sprouting was 1500 units dm$^{-2}$.

Test No. 1 (direct seeding in Table 2) — carbonatites were distributed onto contaminated ground (peat, peat in depression, mineral ground) to create a 5-cm layer and were well watered (10 $\ell$ m$^{-2}$) on 24 June 2011. After that, vermiculite was distributed on the top of the carbonatites to set up a hydroponics layer of 1cm. The mix of grass seeds (114 g m$^{-2}$) was distributed into the vermiculite layer that was re-watered as before (10 $\ell$ m$^{-2}$) and covered by polyethylene film to protect the crops against wind load and water loss due to rapid evaporation. The polyethylene film was filled up with peat for additional protection of the crops from rapid evaporation. After crop germination, the polyethylene film was removed on 1 July 2011.

Test No. 2 (rolled lawn in Table 2) — after two weeks of crop germination in greenhouse conditions, the prepared lawn was divided into pieces of 1 m² square, and the lawn pieces were rolled (Figure 3c) for setting onto the carbonatites layer in the test field on 13 July 2011.

Figure 3. Preparation of rolled lawn: (a) seed and substratum at the initial stage, (b) seed germination and cultivation, and (c) lawn rolling.

## *Substratum Sampling and Chemical Analysis*

Recultivated ground and additives (vermiculite, sand and carbonatites) were sampled prior to their distribution onto the study field. All samples were dried at room temperature (20°C) and put through a sieve with a 1mm grid. After this pretreatment, chemical analysis was carried out as follows: (i) metal elements were extracted from the pre-treated sample using 1M ammonium acetate (i.e. extraction of bio-available forms) (Halonen et al., 1983), and the following elements were determined by atomic absorption spectrometry—K, Na, Ca, Mg, Al, Zn, Fe, Cu, Ni and Mn; (ii) the sieved sample was treated with sulphuric acid and nitric acid, and phosphorous composition was then determined by molybdate colorimetry, and sulfur by turbidimetry with $BaCl_2$; and (iii) 10.0 g of the sieved sample was mixed with 25 ml of distilled water, and then the pH value of the sample was measured by the glass electrode method. Atomic absorption spectrometers (Perkin-Elmer 360, Perkin-Elmer Analyst 800 and AAS-30 Carl-Zeiss Jena) were used for the quantification of metals.

## *Pollution Measurement in the Study Field during 2011-2012*

The snow quality (i.e. wet precipitation of pollutants) was measured in the background field, the test field and the industrial desert in order to assess the pollution load. The precipitated snow cores were sampled in a pit wall using a plastic collector (plastic tube and restricting Plexiglas plate) and plastic bags.

Increments varied from 15-20 cm to whole profile height depending on snow density. Each increment was transferred from the plastic tube to a plastic bag and transported to a chemical laboratory in a frozen state. The samples were melted in a plastic basin in the laboratory, and then the volume of each sample was recorded.

Snow samples were directly analyzed after filtration through filter paper using the following techniques: potentiometry for determining $H^+$, atomic absorption spectrophotometry for determining metal elements, ion chromatography for anion elements, colorimetry for determining P, $P-PO_4^{3-}$, $NH_4^+$ and Si, and the oxidability method using permanganate and bichromate for C determination.

### Plant State Diagnosis and Leaf Diagnostics

Using a square of protjective covering, the density of the vegetative cover, length and mass of overground parts of plants, thickness of the sod cover, and depth of grass roots penetration in contaminated ground were determined during plant state diagnosis.

After grass cover creation, leaves were sampled from the survey plants at the end of the growing season (September 2012). Leaf samples were dried at room temperature, and they were digested with concentrated nitric acid to destroy the matrix and dissolve metals. Metals in a sample solution were determined by atomic absorption spectrometry, phosphorous composition was determined by molybdate colorimetry, and total N was determined by the Kjeldal method.

Leaf diagnostics of the grass cover on the study field were carried out as a comparison with the data from monitoring total content of elements (N, P, K, Ca, Mg, Cu and Ni) in the herbaceous plant Avenella flexuosa L. which is one of the herbaceous plants surviving in the monitoring plots of the authors' research center INEP: the background was monitored at 66°56' 23,1" N and 29°51'22,5"E; the technogenic sparse was monitored at 67°51,08' N and 32°47,49' E.

### Data Processing

Significant levels ($P$ values) were calculated to statistically evaluate the significant difference, and standard deviation values were also calculated to show the dispersion of a set of measured values.

## Results and Discussion

As stated above, this research project is still under way, and it has not yet been concluded. Based on the interim results during 2011–2012, pollution load and biomass diagnosis are presented to give preliminary information about the impact of metals on herbaceous plants.

Many samples were measured in each field, but there is not sufficient space in this chapter to show all the obtained values; without special mention, the average values are shown as representative data: n = 6 in the background field, and n = 3 in the study field.

*Pollution Load*

The data in 2012 show that the pollution level in the study field was much greater than that in the background field. Based on the data in 2012, $P$ values were calculated to statistically evaluate the difference in the pollution level between the background field and the study field, and the obtained values are summarized in Table 3. $H^+$, K, Mg, Mn, P, $PO_4^{3-}$, C, Cd, Pb, Cr and $NO_3^-$ showed no statistical difference between the background field and the study field. Some remarkable points concerning the tendency shown in Table 2 are noted as follows: (i) the precipitation amount of $SO_4^{2-}$ in the study field was over 5.57 times (4406.3 g ha$^{-1}$) greater than that in the background field; (ii) the Cu amount in the study field was over 645 times (572.5 g ha$^{-1}$) greater than that in the background field; (iii) the Ni amount in the study field was over 824 times (685.3 g ha$^{-1}$) greater than that in the background field; (iv) the Co amount in the study field was 808 times (12.3 g ha$^{-1}$) greater than that in the background field; (v) the Fe amount in the study field was 6.6 times (22.7 g ha$^{-1}$) greater than that in the background field; and (vi) the Si amount in the study field was 21 times (111.1 g ha$^{-1}$) greater than that in the background field. From these viewpoints, it is evident that the main factor of environmental degradation (i.e. defoliation and dusting) around the nickel-copper smelter complexes is heavy-metal pollution, and the pollution load is still being imposed on the surrounding vegetation.

The contents of $Na^+$ and $Cl^-$ in the study field were about 6 times (1840,8 g ha$^{-1}$ $Na^+$ and 3184,8 g ha$^{-1}$ $Cl^-$) greater than those in the background field. Since the study field is located near a trunk road, it is reasonable to consider the effects of traffic; anti-freezing agents were used on the snowy and/or icy road surface during winter, so there is a possibility that the saline contents of these agents may have increased $Na^+$ and $Cl^-$ in the study field.

**Table 3. Statistical difference in pollution level between the study field and background field on the basis of the data in 2012**

| $P$ value | Statistical difference | Element |
|---|---|---|
| $P \geq 0.05$ | Not significant | $H^+$, K, Mg, Mn, P, $PO_4^{3-}$, C, Cd, Pb, Cr, $NO_3^-$ |
| $P < 0.05$ | Significant | Ca, Sr |
| $P < 0.01$ | Highly significant | $NH_4^+$, Al |
| $P < 0.001$ | Extremely significant | Cu, Ni, $SO_4^{2-}$, Na, Fe, Si, $Cl^-$, Co |

*Plant Diagnosis*

Figure 4 shows pictures of grass growth and covering rates of grass according to the vegetation season. The 2011 data were collected at the end of the first vegetation season, and the 2012 data were collected at the end of the second vegetation season. Low rates of grass coverage were recorded in the case without the substratum, but the artificial substratum achieved a high rate in 2011 and 2012. It follows from Figure 4 that a carbonatite substratum can keep a high rate of grass coverage. With regard to plant height on the ground, this parameter increased in all tests, but more encouraging results were received where there was application of the carbonatite substratum with direct sowing in the ground depression. The biometric parameters except the plant height are summarized in Table 4; these parameters were obtained after passing two vegetation seasons.

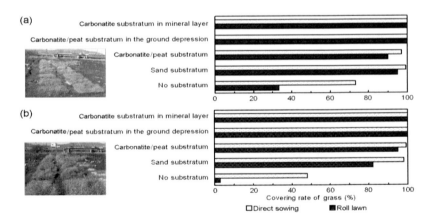

Figure 4. Covering rate of grass (%) as a function of substratum type with grass creation method: (a) the end of the 2011 vegetation season; and (b) the end of the 2012 vegetation season.

**Table 4. Biometric parameters of the grass at the end of the 2012 vegetation season**

| Ground type | Substratum | Grass planting | *Weight ratio | Sod depth, cm | Biomass on the ground, g m$^{-2}$ |
|---|---|---|---|---|---|
| Peat | Not applied | Direct sowing | 2.8±2.2 | 1.5±0.0 | 37 |
| | | Rolled lawn | No data | 1.0±0.0 | 9 |
| | Sand | Direct sowing | 5.3±3.1 | 3.6±0.8 | 256 |
| | | Rolled lawn | 5.7±6.1 | 2.5±0.0 | 246 |
| | Carbonatite | Direct sowing | 2.4±0.9 | 6.9±0.3 | 336 |
| | | Rolled lawn | 1.8±0.7 | 6.8±1.0 | 389 |
| Peat in depression | Carbonatite | Direct sowing | 6.9±0.0 | 6.5±0.0 | 448 |
| | | Rolled lawn | 13.6±0.0 | 6.8±1.0 | 997 |
| Mineral layer | | Direct sowing | 2.7±1.1 | 10.3±3.2 | 784 |
| | | Rolled lawn | 1.7±1.1 | 8.0±1.4 | 761 |

* Weight ratio of above-ground part to underground part.

Obviously, the contaminated ground without any additive retarded root system penetration, resulting in low values of sod characteristics. The best results of the biomass were observed in Variant 2 on peat in depression (997 g m-2). The highest weight ratio of the above-ground part to the underground part was observed in this case too. The best sod characteristics (10.3±3.2 cm) were observed in rolled lawn on the mineral layer.

During the vegetative season of 2012, penetration of the grass root system into the contaminated ground was observed: 1 cm into peat, 2 cm into peat in depression and 3 cm into the mineral layer. There was no root penetration into sand or into contaminated ground without any additive. It could be presupposed that using ameliorant on contaminated ground has a positive influence under artificial layers (vermiculite + carbonatites).

## Element Diagnosis in Leaves

Diagnosis of element effects (deficiency and toxicity) in plants plays an important role in determining plant health. Based on the results of leaf analysis, it is advisable to assess whether supplement fertilization and/or more strict control of metals are necessary. Element contents in the sampled leaves of the test field are shown in Table 5, and these levels are compared with leaf properties in the background zone and those of survival plants in the test field.

**Table 5. Leaf elements in the monitored plants**

| Element | Direct sowing | Rolled lawn | *Background | **Survival grass |
|---------|---------------|-------------|-------------|------------------|
| N       | 15942         | 17688       | 11330       | No data          |
| P       | 608           | 1022        | 1660        | 690              |
| K       | 17286         | 20556       | 12290       | 11490            |
| Ca      | 5388          | 4942        | 1700        | 670              |
| Mg      | 1947          | 2015        | 650         | 660              |
| Cu      | 278           | 204         | 3.9         | 170              |
| Ni      | 316           | 242         | 2.3         | 360              |

*Avenella flexuosa* L. collected in the background zone, and **Avenella flexuosa* L. surviving in the test field in spite of multiple contamination.

The obtained results suggested the compliance of P content in leaves to the viability level (608 mg g$^{-1}$ in direct sowing and 1022 mg kg$^{-1}$ in rolled lawn) in comparison with the survival level of 690 mg kg$^{-1}$ and the background level of 1660 mg kg$^{-1}$.

The levels of K, Ca and Mg in leaves in the study field exceeded survival levels of herbaceous plants. In addition, N, K, and Mg values in leaves in the study field even exceeded background values.

As is seen in Table 5, the Ni contents of leaves in direct sowing and in rolled lawn were 316 and 242 mg kg$^{-1}$ respectively, and the Cu contents were 278 and 204 mg kg$^{-1}$ respectively. These Cu contents are greater than that in the survival plants.

It should be mentioned that no obvious signs of grass degradation such as chlorosis were observed.

This implies that high contents of heavy metal could be partly the result of dry deposition and its impact on the vegetation state is not significant in the incorporation of metals into a plant body.

The leaf contents of N, Ca and Mg were similar in both rolled lawn and direct sowing, but the nutrient contents of P and K in rolled lawn were greater than those in direct sowing; it is therefore considered that the rolled lawn method contributes more to plant health.

## Conclusion

Mining and smelting activities represent a major source of multiple contamination by air pollutants and metal contaminants (Freedman and Hutchison, 1998). This consequently may threaten the health of animals and human beings. Close by the nickel-copper smelters in the Arctic region, the deposition of multiple pollutants has severely damaged the soil and ground vegetation.

Since scientific reports on the responses of field-growing herbaceous plants to the severe local impact of industrial emissions are surprisingly scanty, the authors initiated a field test of grass near a source of multiple pollution (i.e. Ni-Cu smelter) in the sub-arctic region.

The obtained interim results indicate that plant roots proliferated in the contaminated ground up to 3 cm in some cases, and the increment rate of grass was high through two vegetation seasons; however, the influence of freezing and nutrient loss rate are unclear in the study field, so the field study is still continuing.

It is important to quantitatively and qualitatively understand the impact of metal pollution on plants and the resistance mechanism in field conditions from the practical viewpoint. It is hoped that this chapter will contribute to research and development for cleaning and remediating polluted land by means of proper planting.

## Acknowledgments

The project scheme and implementation were organized by Petrozavodsk State University, Polar-Alpine Botanical Garden Institute (PABGI) and the Institute of the North Industrial Ecology Problems (INEP) – Russian Academy of Sciences (RAS).

The authors are grateful to Mrs. I. P. Kremenetskaya (Institute of Chemistry and Technology of Rare Elements and Mineral Raw Materials) for organization of field work and supervision of chemical analyses, Mrs. S. V. Sverchkova (INEP) for supporting chemical analyses of leaves, Ms. G. N. Andreeva (INEP) for supporting chemical analyses of ground and substratum, CERNAS and JST-RISTEX for supporting data processing facilities, and Ms. C. Lentfer for English review.

## References

AMAP Assessment (2006). Acidifying pollutants, arctic haze, and acidification in the Arctic. Oslo: Arctic Monitoring and Assessment Programme.

Bell, J. N. B. and Treshow, M. (2002). *Air pollution and plant life*. Chichester: John Wiley and Sons.

Berlyand, M. E. (1991). *Annual report on ambient air pollution in cities and industrial centres of Soviet Union* (Emission of pollutants 1990). St. Petersburg: Voeikov Main Geophysical Observatory (in Russian).

Freedman, B. and Hutchinson, T. C. (1981). Source of metal and elemental contamination of terrestrial environments. In: N. W. Lepp (ed.), *Effect of heavy metal pollution on plants* (pp. 35-94). London: Applied Science Publishers.

Halonen, O., Tulkki, H. and Derome, J. (1983). Nutrient analyses methods. *Metsantutkimuslaitoksen Tiedonantoja,*121,1-28.

Hassan, I. A. (2004). Interactive effects of salinity and ozone pollution on photosynthesis, stomatal conductance, growth, and assimilate partitioning of wheat (*Triticum aestivum L*). *Photosynthetica*, 42, 111-116.

Ivanova, L. A., Kostina, V. A., Kremenetskaya, M. V. and Inozemtseva, E. S. (2010). Accelerated formation of anti-errosive grass cover on technogenically disturbed territories: Polar region. *MGTU vestnik*, 13 (4/2), 977-983.

Kikuchi, R. and Gorbacheva, T. (2006). Vegetation recovery after environmental damage by metallurgic industry in the Arctic region: Transformation of soil chemistry in restored land. In: C.V. Loeffe (ed.), *Conservation and Recycling of Resources* (pp. 93-118), Hauppauge: Nova Science Publishers.

Kozlov, M. V. and Zvereva, E. L. (2007a). Does impact of point polluters affect growth and reproduction of herbaceous plants? Water, Air and Soil Pollution, 186, 183-194.

Kozlov, M. V. and Zvereva, E. L. (2007b). Industrial barrens: extreme habitats created by non-ferrous metallurgy. *Reviews in Environmental Science and BioTechnology*, 6, 231–259.

Lechowicz, M. J. (1987). Resource allocation by plants under air pollution stress: implications for plant–pest–pathogen interactions. *Botanical Review*, 53, 281–300.

Lokeshwari, H. and G. T. Chandrappa, G. T. (2006). Impact of heavy metal contamination of Bellandur Lake on soil and cultivated vegetation. *Current Science*, 91 (5), 662-627.

Milyaev, V. B. and Yasenskij, A. N. (2004). *Annual report on emissions of pollutants into the atmosphere in cities and regions of the Russian Federation in (2003)*. St. Petersburg: Institute of Ambient Air Protection (in Russian).

Tikkanen, E. and Niemelä, I. (1995). *Kola peninsula pollutants and forest ecosystems in Lapland* (Final report of the Lapland Forest Damage Project). Rovaniemi: Finnish Forest Research Institute.

Treshow, M. and Anderson, F. K. (1989). *Plant stress from air pollution*. Chichester: John Wiley and Sons.

Ping, C. S. (2003). Effects of Heavy Metals on Plants and Resistance Mechanisms. *Environmental Science and Pollution Research*, 10 (4), 256-264.

Qin, P., Tie, B. and Zhou, X. (2000): Effects of Cadmium and Lead in Soil on the Germination and Growth of Rice and Cotton. *Journal of Hunan Agricultural University*, 26 (3), 205-207.

Rigina, O. and Kozlov, M. V. (2000). Pollution impact on sub-Arctic northern taiga forests in the Kola Peninsula, Russia. In: J. L. Innes and J. Oleksyn (eds.), *Forest dynamics in heavily polluted regions* (pp. 37-65), IUFRO Research Series 1. Wallingford: CAB International.

Ryser, P. and Sauder, W. R. (2006). Effects of heavy-metal contaminated soil on growth, phenology and biomass turnover of Hieracium piloselloides. *Environmental Pollution*, 140, 52–61.

Wang, S., Li, J., Shi, S., et al. (2001). Geological disease caused by ecological environment: an example of cancer village in Shanxi Province. *Environmental Protection*, 5 (42/43), 46.

Ward, N. I. (1995). Environmental analytical chemistry. In: F. W. Fifield and P. J. Haines (eds.), *Trace Elements* (pp. 320-328), Glasgow: Blackie Academic and Professional.

Wu, Y., Wang, X. and Liang, R. (1998). Dynamic migration of Cd, Pb, Cu, Zn and As in agricultural ecosystem. *Acta Scientiae Circumstantiae*, 18 (4), 407-414.

Zhang, Y., Cai, T. and Burkard, G. (1999). Research advances on the mechanisms of heavy metal tolerance in plants. *Acta Botanica Sinica*, 41 (5), 453-457.

In: Herbaceous Plants
Editor: Florian Wallner

ISBN: 978-1-62618-729-0
© 2013 Nova Science Publishers, Inc.

*Chapter 5*

# Effects of Abiotic Factors on Herbaceous Plant Community Structure: A Case Study in Southeast Cameroon

*Jacob Willie*[1,2,*], *Eduardo de la Peña*[1], *Nikki Tagg*[2] *and Luc Lens*[1]

[1]Terrestrial Ecology Unit, Department of Biology, Ghent University, Gent, Belgium
[2]Projet Grands Singes, Centre for Research and Conservation, Royal Zoological Society of Antwerp, Antwerp, Belgium

## Abstract

Abiotic factors significantly influence the structure of plant communities, with the effects varying in both space and time. Herbaceous plants belonging to 15 families were monitored in 250 4-m² plots distributed in six habitat types in order to assess the effects of abiotic factors on the abundance of this resource. In each plot, we counted herb

---

[*]Corresponding author: Terrestrial Ecology Unit, Department of Biology, Ghent University, K.L. Ledeganckstraat 35, B-9000 Gent, Belgium; E-mail: Jacob.Willie@UGent.be.

stems and determined the total number of species, the total number of normal stems and the total number of dwarf stems. In addition, we determined soil fertility and other environmental variables. Elevation and soil texture varied, but similar levels of chemical fertility were seen across different habitat types. Herb abundance varied within and between patches, reflecting changes in environmental conditions. Stem biomass was highest in light gaps, and decreased in late successional forests. Light seemed to be the most important factor influencing the abundance of herbs from Marantaceae and Zingiberaceae families only. Despite the hydromorphic nature of the soil in swamps, stem biomass did not exceed that of *terra firma* forests. At the temporal scale, rainfall did not seem to influence stem density as herbaceous plants were available year-round. These results suggest that light might limit the abundance of some herbaceous plants in the study site. However, a long-term investigation is needed to draw firm conclusions on the effects of abiotic factors on herbaceous plant communities in African rain forest.

**Keywords**: Dja reserve, forest understory herbs, plant gradients, stem size; soil fertility

# Introduction

Abiotic variables, among other factors, significantly shape the structure of plant communities (Tilman, 1983; Wright, 1992; Malenky et al., 1993; Crawley, 1997a; Van Andel, 2005; Bonnefille, 2010; Matías et al., 2012). Plant productivity depends on abiotic resources such as light, water and nutrients. Light is needed to catalyze chemical reactions that result in accumulation of plant biomass (Leuschner, 2005; Mooney & Ehleringer, 1997). In turn, these reactions require water, and plants use available water in the air and soil to compensate for the associated loss (Mooney & Ehleringer, 1997). In addition, nitrogen, phosphorus and other nutrients are needed to enhance plant chemical reactions (Fitter, 1997; Mooney & Ehleringer, 1997). For example, under high light conditions, there is a positive correlation between leaf nitrogen content and net photosynthesis (Mooney & Ehleringer, 1997). Tropical herbaceous plant communities are very sensitive to shortages in water, nutrients and light (Wright, 1992).

Across tropical Africa, studies on herbaceous plant community structure have revealed variations in diversity, density and biomass within and across sites (Watts, 1984; Rogers & Williamson, 1987; White et al., 1995; Fay, 1997;

Brugiere & Sakom, 2001; Doran et al., 2002; Ganas et al., 2004; Harrison & Marshall, 2011). It has been suggested that such variations might result from differences in land use history and forest structure and composition, as well as variations in light and soil conditions and other environmental constraints (Brugiere & Sakom, 2001; Baeten et al., 2011). For example, human disturbance of natural habitats may deplete soil resources and negatively affect the recruitment of plant species (Martin et al., 2004). Describing abiotic factors and assessing their relationship with plant communities may therefore provide information on species which are effective indicators of habitat quality and diversity (Moffatt & McLachlan, 2004).

The majority of studies focusing on herbaceous plants across sites of Central Africa have not investigated the influence of environmental factors on the diversity, density and biomass of these plants (e.g. Rogers & Williamson, 1987; Malenky et al., 1993; Furuichi et al., 1997; Brugiere & Sakom, 2001). In some rare cases, the influence of a few abiotic factors has been assessed (e.g. Rogers et al., 1988; but see also Willie et al., 2012). However, the trends that emerge from such studies are often incomplete because the performance of understory plants is a response to the combined effect of a set of environmental factors which vary in magnitude (Ticktin & Nantel, 2004). As a result, other important factors not previously explored can affect the performance of these herbs. A more complete evaluation of the relationship between herbaceous plant communities and environmental parameters is therefore needed. Such investigations may help to highlight the abiotic factors that determine herb availability to potential users such as forest herbivores.

The objective of this study is to assess the influence of abiotic factors on the growth of forest understory herbaceous plants, and provide insights on the causes of gradients in the availability of herbs that are used by forest herbivores. We hypothesize that spatial changes in the magnitude of ecological variables translate to variations in density, diversity and biomass of herbaceous plants. We predict that soil fertility and light and water availability will be the variables that have the greatest influence on herbaceous plant community structure, and will correlate positively with plant abundance and diversity. Assessing density, diversity and growth performance of herbs in relation to abiotic factors might help to describe the environmental features which confer suitability of some species for use by herbivores. In addition, such investigations may provide more clues regarding the environmental characteristics of habitats which are more suitable to herbivores, thus allowing a more accurate assessment of their quality.

## Material and Methods

### Study Site and Species

Data were collected in 'La Belgique' research site of *Projet Grands Singes* (PGS), of the Centre for Research and Conservation (CRC), Royal Zoological Society of Antwerp (RZSA), located between 013°07'–013°11' E and 03°23'–03°27' N. The site is situated in the northern buffer zone of the Dja Biosphere Reserve (Cameroon), and is located in the transition zone between the semi-deciduous forests of Equatorial Guinea and the evergreen forests of the Congo basin (Letouzey, 1985). The climate is equatorial and humid and is characterized by seasonal rainfall. During a two-year period (April 2009–March 2011), average rainfall was 1637.9 ± SD 105.1 mm, and mean minimum and maximum daily temperatures ranged between 19.5 ± SD 1.3°C and 26.3 ± SD 2.4°C. The study subjects were herbaceous species belonging to 15 families, namely Araceae, Aspleniaceae, Balanophoraceae, Commelinaceae, Costaceae, Cyperaceae, Marantaceae, Melastomataceae, Poaceae, Pteridaceae, Rubiaceae, Selaginellaceae, Thelypteridaceae, Urticaceae and Zingiberaceae. It is noteworthy that Marantaceae and Zingiberaceae density in the study site is about 3 stems/m² (Willie et al., 2012), whereas overall herb density is estimated at 6 stems/m² (Willie et al., in prep.), therefore meaning that herbs from these two families are an important component of the herbaceous layer in the study site.

### Habitat Types

In line with previous vegetation classifications in the area (e.g. Nguenang & Dupain, 2002; Dupain et al., 2004), we distinguished six habitat types: 1) Near primary forest (NPF), where large tree species of height > 30 m predominate (e.g. *Polyalthia suaveolens*, *Omphalocarpum procerum*, *Uapaca* spp. and *Piptadeniastrum africanum*), and there is little undergrowth and a closed canopy; 2) Old secondary forest (OSF), with dominant canopy trees of height 25–30 m (e.g. *Terminalia superba*), a more dense understory than NPF, and a discontinuous canopy layer; 3) Young secondary forest (YSF), characterized by a canopy height of < 25 m dominated by early successional trees (e.g. *Myrianthus arboreus*, *Tabernaemontana crassa*), and a relatively dense undergrowth; 4) Light gaps (LG), with completely open canopies resulting from elephant activity or tree and branch fall; 5) Swamps (SW), with high densities of *Raphia* spp., rare (< 5%) raphia-free open areas (clearings), and a hydromorphic soil; and 6) Riparian forest (RF), growing in the transition zone between SW and other habitat types, with a highly heterogeneous floristic composition comprising species from all habitat types. NPF, OSF, YSF and LG are referred to collectively as *terrafirma* habitats. SW and RF are (periodically) flooded habitats.

## Sampling Design and Characterization of Plots

Stems of all ground-rooted herb species were surveyed in 250 2 × 2 plots placed along 10 transects. Each transect was 6 km long and set at a bearing of 45°. Along each transect, 25 plots were set 250 m apart, at the right side of transects, and at a perpendicular distance of 5 m. Habitat type for each plot was noted. Percentage canopy cover above each plot was visually described (Loya & Jules, 2008) by assigning cover classes and light scores as follows: closed (0), half-open (50) and open (100). Soil humidity scores were determined in a 100–300 scale (100 = *terra firma* habitats on well-drained soils; 200 = riparian forest in the transition zone between *terra firma* habitats and swamps; 300 = swamps on hydromorphic soils). Geographic coordinates and elevation for each plot were also recorded using a GPS Map60cx. Soil samples were collected in 50 plots (7–10 randomly selected plots per habitat type) in 10 × 10 × 15 cm (depth) volumes and analyzed in the lab to determine the pH and the content of organic matter, sand, clay and nutrients (e.g. nitrogen, phosphorus, potassium).

To assess the spatial structure of the herbaceous plant community, we identified and counted herb stems in 4-m² contiguous square plots along a 1.5-km transect traversing all habitat types at a bearing of 45°. Average stem density for each habitat patch encountered along the transect was calculated, and results were graphically portrayed to highlight the patterns of variation.

## Phenological Monitoring of Herbaceous Plants

Herb stems were monitored in all plots (250) along the 10 6-km transects. In each plot, the total number of herb stems and species were determined. Each stem was examined, and only old stems were classified as "dwarf" or "normal" because they had already completed their developmental cycle. Classification was based on size, which affects vegetative propagation (Ticktin

& Nantel, 2004). Old stems were distinguished by signs of age, such as the occurrence of many yellow or brown leaves (entirely or partially), sometimes with holes and a dull color. Dwarf and normal stem dimensions were mutually exclusive. In all cases, dwarf stems were less than half the potential plant height. We collected and weighed ten stems of each species (five normal and five dwarf), and only one overlap in weight was found out of 600 measures. Each stem was assessed based on these chosen limits using a small decameter (with millimeter precision). For each species, we recorded the following information: 1) total number of stems; 2) number of dwarf stems; 3) number of normal stems; 4) number of stems with flowers; 5) number of stems with fruits and 6) number of growing shoots. These data alongside rainfall, humidity and temperature data were collected each month, from August 2011 to July 2012.

## Effects of Environmental Factors on Herbaceous Plants

To assess the effects of abiotic factors on herbaceous plant community structure, we performed a nonmetric multidimensional scaling (NMDS) ordination in R using the Euclidian distance on log transformed abundances. The data matrix was composed of herb and environmental data collected on a subset of 50 randomly-chosen plots (7–10 per habitat type). Analyses were done using the Euclidian distance on log transformed abundances. A preliminary principal component analysis (PCA) was performed in XLSTAT in order to assess autocorrelations among environmental variables (Figure1). For final analyses, four non correlated variables were chosen. Component scores for each plot in the original data matrix were calculated, and correlations between dependent variables and principal axes were assessed using Spearman tests of correlation.

## Statistical Analyses

Nonparametric statistics were used as data did not meet the assumptions of normality. Median tests (two-tailed) were used for global comparisons of habitat types, and two-sample Kolmogorov-Smirnov tests (two-tailed) were used for pairwise comparisons. We did not apply the Bonferonni correction for pairwise comparisons as sample sizes were too small ($7 \leq N \leq 60$ in most cases; Garamszegi, 2006). Proportions were compared using Chi-squared tests. Statistical analyses were run in SPSS.

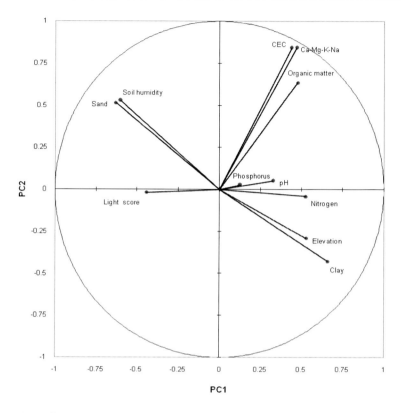

Figure 1. Preliminary principal component analysis (PCA) of environmental variables. Highly autocorrelated variables are more close to each other.

# Results

## Characteristics of Habitat Types

There was little variation in plot elevation across habitat types, though differences in medians were statistically significant (Table 1). No significant difference in elevation was found among *terra firma* habitat plots or among flooded habitat plots; whereas plots in all *terra firma* habitats, except in young secondary forest, had significantly different elevations compared to plots in swamps (two-sample Kolmogorov-Smirnov tests, $P < 0.01$ in all cases). Sand and clay content of the soil (proxies for soil texture) globally differed across habitat types (Table 1). However, no significant difference was found among *terra firma* habitats and among flooded habitats, and all *terra firma* habitat soils had significantly different sand and clay content compared to swamps (two-sample Kolmogorov-Smirnov tests, $P< 0.05$ in all cases). Levels of soil fertility, estimated using measures of pH, organic matter, cation exchange capacity (CEC) and availability of phosphorus and nitrogen did not significantly differ across habitat types (Table 1).

## Spatial Variation in Herb Availability

Along the 1.5 km transect, a mosaic of small-sized patches of various habitat types was encountered. The highest herb densities occurred in light gaps, young secondary forest and swamp patches. Furthermore, there was considerable variation in herb density among patches within each habitat type (Figures 2 and 3).

## Abiotic Factors and Herb Community Structure

A preliminary synthesis of environmental parameters using a principal component analysis (PCA) indicated high levels of autocorrelation among variables (Figure 1). To facilitate the interpretation of results, we clustered correlated variables into four groups. Each group was represented by the most meaningful variable, namely clay, organic matter, canopy cover and soil humidity. The NMDS ordination of the herbaceous plant community structure in 50 plots resulted in a two dimensional solution, and the final stress was 0.24. As shown in Figure 4, no clear relationship between predictors (clay, organic matter, light score and soil humidity) and dependent variables (stem density, stem biomass and species diversity) was detected. Plots from light gaps, riparian forest and swamps were separated from the others along the second axis. The first axis was roughly related to dependent variables while the second axis was mainly related to the predictors. Correlations with ordination scores were significant only between stem density and axis 1 ($r_s = -0.31$; $P = 0.027$). To corroborate these results, we assessed the relationship between measured environmental variables, namely sand, clay, pH, organic matter, nitrogen, phosphorus, CEC, light score, soil humidity and elevation, and dependent variables using the Spearman test of correlation: correlation was significant only between species diversity and elevation ($r_s = 0.31$; $P = 0.029$).

## Table 1. Soil characteristics by habitat type

| Variable | Near primary forest | Old secondary forest | Young secondary forest | Light gaps | Riparian forest | Swamps | Significant differences |
|---|---|---|---|---|---|---|---|
| Number of plots | 8 | 7 | 9 | 8 | 8 | 10 | |
| Elevation (m) | 675.25 (17.65) | 682.86 (26.30) | 674.78 (24.57) | 674.63 (12.19) | 663.13 (19.25) | 650.00 (8.00) | ** |
| Sand content (%) | 11.75 (2.05) | 10.43 (0.98) | 10.56 (2.74) | 12.50 (1.60) | 25.13 (15.97) | 24.40 (8.69) | * |
| Clay content (%) | 66.25 (3.49) | 66.00 (3.27) | 66.00 (4.15) | 67.25 (1.58) | 53.00 (15.29) | 52.70 (10.30) | * |
| pH | 4.35 (0.43) | 4.14 (0.62) | 4.54 (0.41) | 4.10 (0.30) | 4.08 (0.17) | 4.13 (0.26) | ns |
| Organic matter (%) | 5.71 (0.93) | 5.80 (1.44) | 5.56 (1.35) | 4.75 (1.67) | 6.55 (1.51) | 5.22 (0.82) | ns |
| Cation exchange capacity (milliequivalents/100g) | 3.07 (0.66) | 3.11 (0.57) | 3.11 (0.68) | 2.63 (0.61) | 3.29 (0.26) | 3.08 (0.33) | ns |
| Assimilable phosphorus (mg/kg) | 3.83 (1.01) | 3.05 (0.73) | 3.65 (0.90) | 3.78 (0.87) | 3.93 (1.73) | 3.45 (0.55) | ns |
| Total nitrogen (g/kg) | 1.99 (0.77) | 1.81 (0.31) | 1.94 (0.40) | 1.75 (0.63) | 1.76 (0.44) | 1.59 (0.42) | ns |

Displayed figures for all measured parameters are average values and corresponding standard deviations (in parentheses); soil parameters were measured in 50 plots selected in all habitat types. All global comparisons were done using Median tests; df = 5 in all cases; ns: non significant; **significant at $P < 0.01$; *significant at $P < 0.05$.

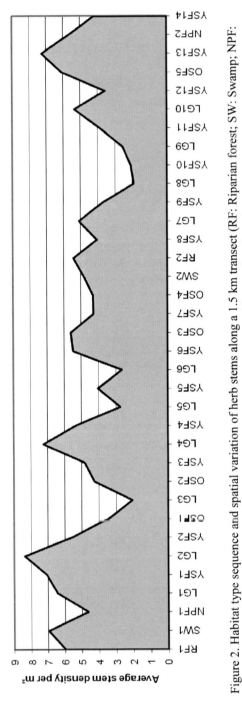

Figure 2. Habitat type sequence and spatial variation of herb stems along a 1.5 km transect (RF: Riparian forest; SW: Swamp; NPF: Near primary forest; LG: Light gap; YSF: Young secondary forest; OSF: Old secondary forest).

Effects of Abiotic Factors on Herbaceous Plant ... 123

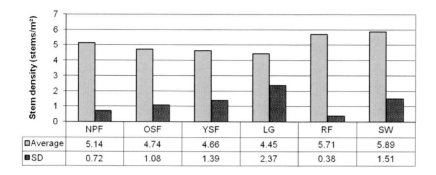

Figure 3. Average values of herb stem density among habitat patches along the 1.5 km transect (NPF: Near primary forest; OSF: Old secondary forest; YSF: Young secondary forest; LG: Light gaps; RF ;Riparian forest; SW: Swamps).

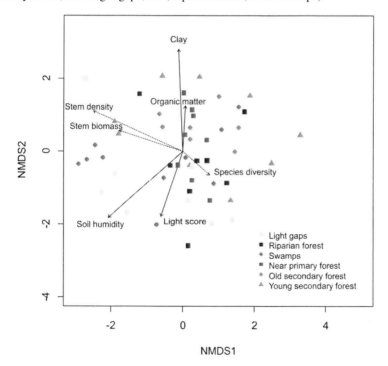

Figure 4. Nonmetric Multidimensional Scaling ordination showing the relationship between habitat type, abiotic factors and herbaceous plant community structure. The symbols represent plots in each habitat type grouped relative to the floristic similarity. The arrows indicate strength and direction of correlations among habitat characteristics and ordination scores ($r^2$ ranged between 0.02 and 0.2).

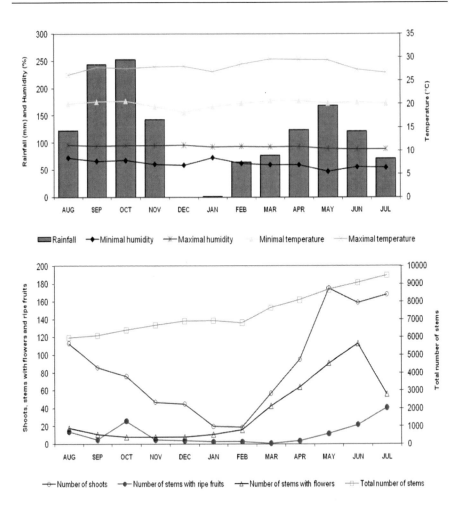

Figure 5. Climatic variables and phenological monitoring of herbs. Data were collected monthly in 250 4-m² plots.

At the temporal scale, abiotic factors such as rainfall, temperature and air humidity, as well as the phenology and dynamics of herbaceous plants showed variations throughout the year (Figure 5).

Rainy seasons spanned from August–November and April–June, and dry season periods occurred in July and from December–March.

Highest and lowest temperatures were noted in March and April, and in August and December, respectively. Air humidity peaked in August, and the minimal level was observed in May.

Small variations in the total number of stems inventoried were noted, with the highest values in July and lowest in value in August. The number of stems recorded showed little variation from August–February, and significantly increased from March–July.

From May–July, high numbers of herb shoots were counted, and low numbers were recorded in January and February. Maximal numbers of herb stems with flowers were observed from March–July, and minimal numbers from August–February. Numbers of herb stems with ripe fruit were high July and October, and minimal values were recorded from January–April.

We used a Spearman test of correlation to assess possible relationships between herbaceous plant density and climatic variables. No significant correlation was detected between the total number of stems recorded each month and rainfall ($r_s = -0.315$; $P = 0.319$; $N = 12$) or average temperature ($r_s = 0.266$; $P = 0.404$; $N = 12$). However, there was a negative correlation with average air humidity ($r_s = -0.853$; $P = 0.0004$; $N = 12$).

Similarly, the number of shoots and the number of stems with flowers and ripe fruits did not show clear relationships with climatic variables, and a significant correlation was detected only between the number of stems with flowers and average air humidity ($r_s = -0.649$; $P = 0.022$; $N = 12$).

## Herb Size Categories

There was a significant difference in the proportion of dwarf and normal stems between habitat types (Chi-squared test: $\chi^2 = 119.6$, df = 5, $P < 0.0001$). These proportions significantly differed while only considering herbs from Marantaceae and Zingiberaceae families (Chi-squared test: $\chi^2 = 89.23$, df = 5, $P < 0.0001$; Table 2).

The global chi-square value was partitioned to investigate specific differences. For all herbs, detailed results indicated significantly different proportions only in young secondary forest and light gaps ($\chi^2 = 22.14$ and 92.77, df = 1, $P < 0.001$ in all cases).

For Marantaceae and Zingiberaceae, significant differences occurred in young secondary forest, light gaps and riparian forest (Chi-squared test: $\chi^2 = 9.65$; 38.07 and 40.38, respectively; df = 1 in all cases; $P < 0.001$ in all cases).

## Table 2. Frequencies of Herb Stem Categories in Different Habitat Types

| Stem size | Near primary forest | Old secondary forest | Young secondary forest | Lightgaps | Riparian forest | Swamps |
|---|---|---|---|---|---|---|
| *Marantaceae and Zingiberaceae only* | | | | | | |
| Dwarf | 152 | 415 | 257 | 66 | 55 | 289 |
| Normal | 154 | 461 | 363 | 184 | 10 | 287 |
| % dwarf | 49.67 | 47.37 | 41.45 | 26.40 | 84.62 | 50.17 |
| *All herb species* | | | | | | |
| Dwarf | 300 | 696 | 494 | 116 | 259 | 361 |
| Normal | 392 | 897 | 559 | 249 | 114 | 384 |
| % dwarf | 43.35 | 43.69 | 46.91 | 31.78 | 69.44 | 48.46 |

## Discussion

Flooded habitats were located in small depressions, and this resulted in lower elevations as compared to *terra firma* habitats (Table 1). Although soil texture varied between these major groups, the chemical characteristics of soils were similar in all habitat types, as also observed by Peh (2009) inside the Dja Reserve (Cameroon). This may be an indication that litter accumulation and decomposition rates and litter nutrient release do not significantly vary across habitat types, despite textural and hydromorphic differences. However, values of standard deviation of the measured environmental variables were sometimes high (Table 1), highlighting spatial variability in the magnitude of these factors. Furthermore, in the study site, the amount of light varies within and between habitat types (Willie et al., 2012). As a result, spatial variations in stem density between and within habitat types (Figures 2 and 3) probably resulted from this environmental variability, especially as tropical terrestrial herbs are very sensitive to such variations (Wright, 1992). This is consistent with the fact that plant germination, recruitment and mortality across forest patches are likely to be affected by variability in environmental conditions (Martinez-Ramos et al., 1989).Our results highlight that while idiosyncratic differences between habitat types or sites may explain observed differences in levels of ecological factors (Brugiere & Sakom, 2001; Loya & Jules, 2008; Baeten et al., 2011), within-habitat variation in abiotic factors results in uneven biological patterns.

However, no strong correlation was detected between abiotic variables and herb abundance and diversity (Figure 4). Hydromorphic and textural differences between flooded and *terra firma* habitats did not seem to result in variations in herbaceous plant community structure. Moreover, no relationship existed between light score and herbaceous plant density, diversity and biomass. This may be an indication that these abiotic factors did not limit the growth of the herbaceous plant community. However, while only considering herbs from Marantaceae and Zingiberaceae families, the proportion of dwarf stems was lowest in light gaps, and consistently increased from young secondary to near primary forest (Table 2). In the study site, the amount of light drastically decreases as the forest progresses from early to late successional stage (Willie et al., 2012). In addition, as shown by the analyses, variations in soil fertility levels across habitats were not significant. Therefore, it is possible that light strongly limits the growth and development of Marantaceae and Zingiberaceae herb species in the study site. It is shown that the stem density of herbs from these two families peaks in light gaps, with minimal values in near primary forest (Willie et al., 2012). These trends are consistent with the observations of White et al. (1995) who found differences in growth formsof *Haumania liebrechtsiana* between Marantaceae and mature forest, with shorter stems occurring in the latter habitat type where shade conditions predominate in the understory. Stems of other plant life-forms such as saplings have displayed similar negative correlations between 'slenderness' and light availability (Van Breugel et al., 2012). In the ecological literature, it is well established that light is a major determinant of plant growth (Mooney & Ehleringer, 1997; Leuschner, 2005). Our analyses, however, did not reveal a clear effect while considering all herb species probably because some of the studied herb species are generalists (Willie et al., in prep.), meaning that they can thrive in a range of habitat conditions. For example ferns, an important component of the herbaceous layer in the study site, can adapt and persist in shade conditions (Crawley, 1997b). Hence, we can suggest that although light had no major influence on the herbaceous plant community as a whole, some species appeared to be more dependent.

Water availability did not seem to limit herbaceous plant growth in the study site. One would have expected the highest stem density and biomass in flooded habitats if water was a liming factor to herbs in *terra firma* habitats. Stem density seemed to be highest in swamp patches (Figure3), but differences in stem density between this habitat and others are not significant (Willie et al., in prep). In addition, riparian forest and swamps had high proportions of dwarf stems compared to other habitats (Table 2), implying that other factors may be important. Moreover, at the temporal scale, no clear pattern was detected between rainfall and density of herbaceous plants (Figure 5). This result must be interpreted with caution because between consecutive surveys, other factors such as herbivory, trampling by animals and humans and other biotic influences might have removed some stems, including shoots, fruits and flowers, thus biasing the results. Nevertheless, these results suggest that herbs are available year round. It has been noted that tropical herbaceous plants are very sensitive to water shortage which can cause mortality and inhibit the germination of some species (Wright, 1992; Crawley, 1997b). In spite of this, the herbaceous plant community of the study site did not seem to be affected by seasonal rainfall as well as variation in temperature. However, further investigation is needed to draw firm conclusions.

## Conclusion

Soil nutrients for plant growth are randomly distributed across habitats. Contrary to our predictions, abiotic resources were not limiting factors to the herbaceous plant community of the study site as a whole, though light seemed to be the most important factor influencing the abundance of a number of species. The effect of climatic variables was unclear, although it is suggested that rainfall is a major limiting factor to plant growth in the tropics (Bonnefille, 2010), and that herbaceous plants are very sensitive to climatic stress (Wright, 1992). Further investigations over a longer period are needed to provide a more accurate assessment of the impact of climatic variability on herbaceous plant communities in African forests.

## Acknowledgments

Financial and logistic support were provided by the Centre for Research and Conservation of the Royal Zoological Society of Antwerp (Belgium), core-funded by the Flemish Government. We thank the Ministry of Forestry and Wildlife and the Ministry of Scientific Research and Innovation, Cameroon, for permission to carry out this research. We deeply appreciate the assistance of John Carlos Nguinlong, Luc Tedonzong and Charles Yem Bamo during data collection and processing. We acknowledge with much gratitude the help of Jean Tongo and his team during field work. Thanks are also extended to Dr Antoine Mvondo Ze for soil sample analyses, and to Dr Lander Baeten for assistance in multivariate analyses. Constructive comments on the manuscript were received from a number of reviewers.

## References

Baeten, L., Verstraeten, G., De Frenne, P., Vanhellemont, M., Wuyts, K., Hermy, M. & Verheyen, K. (2011). Former land use affects the nitrogen and phosphorus concentrations and biomass of forest herbs. *Plant Ecology* 212, 901–909.

Bonnefille, R. (2010). Cenozoic vegetation, climate changes and hominid evolution in tropical Africa. Global and Planetary Change 72, 390–411.

Brugiere, D. & Sakom, D. (2001). Population density and nesting behaviour of lowland gorillas (*Gorilla gorilla gorilla*) in the Ngotto forest, Central African Republic. *Zoological Journal London* 255, 251–259.

Crawley, M.J. (1997a). The structure of plant communities. In M.J. Crawley (Ed.), *Plant ecology* (second edition, pp. 475–531). Oxford, OX2 0EL: Blackwell Science.

Crawley, M.J. (1997b). Life history and environment. In M.J. Crawley (Ed.), *Plant ecology* (second edition, pp. 73–131). Oxford, OX2 0EL: Blackwell Science.

Doran, D.M., McNeilage, A., Greer, D., Bocian, C., Mehlman, P. & Shah, N. (2002). Western lowland gorilla diet and resource availability: New evidence, cross-Site comparisons, and reflections on indirect sampling methods. *American Journal of Primatology* 58, 91–116.

Dupain, J., Guislain, P., Nguenang, G.M., De Vleeschouwer, K. & Van Elsacker, L. (2004). High chimpanzee and gorilla densities in a non-protected area of the northern periphery of the Dja Faunal Reserve, Cameroon. *Oryx* 38, 1–8.

Fay, J.M. (1997). The ecology, social organization, populations, habitat and history of the western lowland gorilla (*Gorilla gorilla gorilla* Savage and Wyman 1847). Ph. D. thesis. St. Louis, Washington University.

Fitter, A. (1997). Nutrient acquisition. In M.J. Crawley (Ed.), Plant ecology (second edition, pp. 51–72). Oxford, OX2 0EL: Blackwell Science.

Furuichi, T., Inagaki, H. & Angoue-Ovono, S. (1997). Population density of chimpanzees and gorillas in the Petit Loango Reserve, Gabon: employing a new method to distinguish between nests of the two species. *International Journal of Primatology* 18, 1029–1046.

Ganas, J., Robbins, M.M., Nkurunungi, J.B., Kaplin, B.A. & McNeilage, A. (2004). Dietary variability of mountain gorillas in Bwindi Impenetrable National Park, Uganda. *International Journal of Primatology* 25, 1043–1072.

Garamszegi, L. (2006). Comparing effect sizes across variables: generalization without the need for Bonferroni correction. *Behavioral ecology* 17, 682–687.

Harrison, M.E. & Marshall, A.J. (2011). Strategies for the use of fallback foods in apes. *International Journal of Primatology* 32, 531–565.

Letouzey, R. (1985). Notice de la carte phytogéographique du Cameroun au 1/500 000. Domaine de forêt dense humide toujours verte. *Toulouse*, Institut de la Carte Internationale de la Végétation.

Leuschner, C. (2005). Vegetation and ecosystems. In E. van der Maarel (Ed.), *Vegetation ecology* (first edition, pp. 85–105). Oxford, OX4 1JF: Blackwell Publishing.

Loya, D.T. & Jules, E.S. (2008). Use of species richness estimators improves evaluation of understory plant response to logging: a study of redwood forests. *Plant Ecology* 194, 179–194.

Malenky, R., Wrangham R., Chapman, C. & Vineberg, E. (1993). Measuring chimpanzee food abundance. *Tropics* 2, 231–244.

Martin, P.H., Sherman, R.E. & Fahey, T.J. (2004). Forty years of tropical forest recovery from agriculture: structure and floristics of secondary and old-growth riparian forests in the Dominican Republic. *Biotropica* 36, 297–317.

Martinez-Ramos, M., Alvarez-Buylla, E. & Sarukhan, J. (1989). Tree demography and gap dynamics in a tropical rain forest. *Ecology* 70, 555–558.

Matías, L. Quero, J.L., Zamora, R. & Castro, J. (2012). Evidence for plant traits driving specific drought resistance. A community field experiment. *Environmental and Experimental Botany* 81, 55–61.

Moffatt, S.F. & McLachlan, S.M. (2004). Understorey indicators of disturbance for riparian forests along an urban–rural gradient in Manitoba. *Ecological Indicators* 4, 1–16.

Mooney, H.A. & Ehleringer, J.R. (1997). Photosynthesis. In M.J. Crawley (Ed.), *Plant ecology* (second edition, pp. 1–27). Oxford, OX2 0EL: Blackwell Science.

Nguenang, G. M. & Dupain, J. (2002). Typologie et description morpho-structurale de la mosaïque forestière du Dja : Cas du site d'étude sur la socio-écologie des grands singes dans les villages Malen V, Doumo-pierre et Mimpala (Est-Cameroun). Anvers, SRZA.

Peh, K.S.H (2009). The relationship between species diversity and ecosystem function in low- and high-diversity tropical African forests. PhD Thesis. Leeds, University of Leeds.

Rogers, M.E. & Williamson, E.A. (1987). Density of herbaceous plants eaten by gorillas in Gabon: Some preliminary data. *Biotropica*19, 278–281.

Rogers, M.E., Williamson, E.A., Tutin, C.E.G. & Fernandez, M. (1988). Effects of the dryseason on gorilla diet in Gabon. *Primates* 22, 25–33.

Ticktin, T. & Nantel, P. (2004). Dynamics of harvested populations of the tropical understory herb *Aechmea magdalenae* in old-growth versus secondary forests. *Biological Conservation* 120, 461–470.

Tilman, D. (1983). Plant succession and gopher disturbance along an experimental gradient. *Oecologia* 60, 285 –292.

Van Andel, J. (2005). Species interactions structuring plant communities. In E. van der Maarel (Ed.), *Vegetation ecology* (first edition, pp. 238–264). Oxford, OX4 1JF: Blackwell Publishing.

Van Breugel, M., Van Breugel, P. Jansen, P.A., Martínez-Ramos, M. & Bongers, F. (2012). The relative importance of above- versus belowground competition for tree growth during early succession of a tropical moist forest. *Plant Ecology* 213, 25–34.

Watts, D.P. (1984). Composition and variability of mountain gorilla diets in the central virungas. *American Journal of Primatology* 7, 323–356.

White, L.J.T., Rogers, M.E., Tutin, C.E.G., Williamson, E. & Fernandez, M. (1995). Herbaceous vegetation in different forest types in the Lopé Forest Reserve, Gabon: implications for keystone food availability. *African Journal of Ecology 33*, 124–141.

Willie, J. Petre, C.A., Tagg, N. & Lens, L. (2012). Density of herbaceous plants and distribution of western gorillas in different habitat types in south-east Cameroon. African Journal of Ecology. doi: 10.1111/aje.12014.

Wright, S.J. (1992). Seasonal drought, soil fertility and the species density of tropical forest plant communities. *Trends in Ecology and Evolution* 6, 159–152.

In: Herbaceous Plants
Editor: Florian Wallner

ISBN: 978-1-62618-729-0
© 2013 Nova Science Publishers, Inc.

Chapter 6

# Environmental Performance of Three Novel Opportunity Biofuels: Poplar, Brassica and Cassava during Fixed Bed Combustion

*Maryori Díaz-Ramírez[1,2]\*, Christoffer Boman[3], Fernando Sebastián[2], Javier Royo[1], Shaojun Xiong[4] and Dan Boström[3]*

[1]Department of Mechanical Engineering, University of Zaragoza, Zaragoza, Spain
[2]Centre of Research for Energy Resources and Consumption, CIRCE Foundation, Zaragoza, Spain
[3]Thermochemical Energy Conversion Laboratory, Department of Applied Physics and Electronics, Umeå University, Umeå, Sweden
[4]Biomass Technology and Chemistry, Swedish University of Agricultural Sciences, Umeå, Sweden

---

\* E-mail address: mdiaz@unizar.es, phone number: +34-976-76-25-82.

## Abstract

In the last few decades several types of solid biofuels have been proposed as possible sources for heat generation because of growing concerns about environmental pollution, and future fossil fuel supply uncertainties. Among other biomass assortments, short rotation coppice and herbaceous species have been considered. An important aspect to be evaluated to enable a sustainable introduction of such novel fuels is related to their environmental performance during combustion. In this work, three fuel types; one herbaceous energy crop and one short rotation coppice (both cultivated and pelletized in Spain), together with one agricultural residue (cultivated in China) have been assessed in terms of their emission levels of gases (CO and $NO_X$) and particulate matter. The experiments showed that combustion of the fuels was attained under an acceptable level of CO emissions.

However, concentration of $NO_X$ was rather high, but perhaps more important, a considerably high formation of fine particle emissions was observed. Consequently, the incorporation of primary or secondary particle precipitating reduction measures might be needed. In addition, the high ash content in these fuels can severely deteriorate the combustion performance and reliability. Thus, specially designed burners/grate units are therefore needed if a utilization of these fuels in small and medium scale combustion systems seeks to be feasible. Although the applicability of introducing this kind of biofuels to the residential heating sector perhaps seems to be rather limited, it should not always be rejected. Nevertheless, technology improvements would have to be considered to manage the current limitations.

## Introduction

Stemwood has generally been preferred as raw material in pelletized biofuels used for heating supply because it is considered to have preferable combustion properties compared to other kind of pelletized biofuels. Nevertheless, availability limitations of these forest sources, and growing economic and environmental pressures related to fossil fuels have contributed to address efforts on seeking for novel opportunity biofuels, such as dedicated energy crops and agricultural sources [1-4]. These novel fuels are typically characterized by a high variability of their physical and chemical properties, which may lead to undesired effects during combustion [5-9].

For heating applications, grate conversion has been the most extensively applied fixed bed technology to direct combustion of solid biofuels. The foremost advantages for its commercial applicability in the heating sector are its rather low investments and operating costs in comparison to other combustion technologies. Furthermore, this type of technology has also been traditionally preferred to manage heterogeneous fuels. Nevertheless, rather few experiences have so far been carried out with dedicated energy crops.

An important aspect to be evaluated to enable a sustainable introduction of these fuels is related to their environmental performance during combustion. Pollutants typically associated to biomass combustion systems are classified as gases, such as CO and $NO_X$ emissions, particulate matter emissions and ash residues, mainly caused by both organic and inorganic components in the biofuels.

To gain more knowledge of the environmental performance of some of these novel biofuels, two varieties of energy crops cultivated in Spain, (i.e., a herbaceous energy crop, *Brassica carinata* and a short rotation coppice, *Populus sp.*), together with cassava stems (i.e., *Manihot esculenta*), one agricultural residue cultivated in China, were tested as pellets in a residential fixed bed grate conversion appliance with a nominal heat output of 25 $kW_{th}$. The main purpose of the combustion tests was to evaluate the environmental performance of the three fuel types with regard to CO, $NO_X$ and particulate matter emissions. Experimental data have been compared to the European restrictions to gain knowledge concerning potential deviations.

Finally, possibilities to enhance the use of these novel fuels with regard to their experimentally determined combustion and environmental performance in this kind of small scale heating systems are also discussed.

# 1. Combustion Experiences in a Fixed-Grate Burner

## 1.1. General Fuel Characteristics

A description of the fuel properties for the three biofuel types studied brassica, poplar and cassava fuels directly linked to their environmental performance is presented in this section. Further information concerning the three biofuel types can be found in previous research [10].

Lower heating values were rather similar among the fuels and around 17 MJ/kg, dry basis (d.b.). A more significant variability was found for the ash content (incinerated at 550°C). It varied much more significantly between the fuels and in all the cases it was higher compared to the ash content specified for standard softwood pellets. According to the European standard EN 303-5:1999, high quality standardized woody pellets should have an ash content lower than 0.5 weight percent, dry basis (wt% d.b.) [11]. For the herbaceous fuel brassica, the ash content was 8.5 wt% d.b., which was almost twice the one for cassava (i.e., 4.5 wt% d.b.) and three times the corresponding for poplar (i.e., 2.8 wt% d.b.). Based on these results, the fuels can be generally grouped as "ash-rich" fuels. A high ash content in the fuels might affect their combustibility [7, 9]. During combustion, ash may hinder an effective contact between air and fuel and, therefore, adjustments on air supply/distribution or burner/grate design changes to achieve longer residence times can be needed to manage the fuel properties.

Furthermore, fuels exhibited high differences with respect to the volatile matter/carbon fixed (VM/FC) ratio determined by data obtained from the proximate analysis. Concerning this ratio, poplar yielded the best results whereas the worst was found for brassica. The VM/FC ratio for poplar was approximately 22% higher than the lowest value determined for brassica. These results might also affect combustibility of the fuels. For instance, primary air proportion required for particle combustion might be rather different among fuels. Results from the ultimate analysis basically demanded special attention on the N-content because of its influence on $NO_X$ emissions. Cassava and brassica fuels depicted a relatively similar value, 1.4 and 1.8 wt% d.b., respectively [10]. This condition considerably exceeded the N-content for poplar (0.4 wt% d.b.). It was approximately 5 times higher [10].

Finally, concerning the inorganic elements, concentrations vary over quite a broad range. In general terms, main ash components among the fuels were Si, Ca, K, Mg, P, Cl and S. All these elements may interact and lead to problematic ash related effects.

Additional information about the ash components is out of the scope of this research and more detailed information concerning the ash transformations during combustion of these fuels can be found elsewhere [10]. Still, the combustion performance aspects related to the high ash content in these biofuels are briefly discussed here.

## 1.2. Conversion System and Emission Measurements

Main characteristics of the conversion system used in this work are mentioned in this section. A detailed explanation of this system is given in a previous work [10]. A schematic view over the experimental setup is presented in Figure 1.

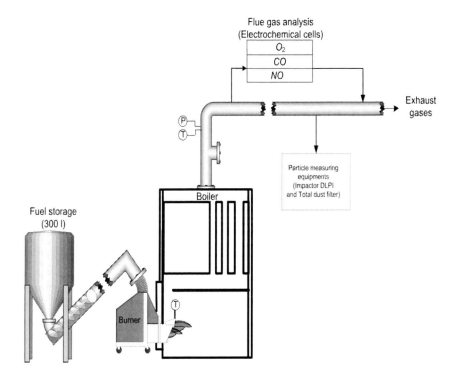

Figure 1. Schematic illustration of the experimental setup.

The experiments were performed in a reference boiler which also is used in the Swedish national certification system (P-marking). Besides the integrated heat exchanger, the boiler walls of this system are also water jacketed. This conversion unit was designed and commercially used for residential heating based on wood pellet combustion with a nominal heat output of up to approximately 25 $kW_{th}$.

The burner used in this work is top-fed (fuel supply) and constructed like a small-grate burner according to the main principles required for pellet combustion. It is similar to a pot and has only one grate where primary air is underfeed, whereas secondary air is distributed inside the pot by several nozzles placed on the sides. Air distribution cannot be adjusted by the user since primary and secondary air flows are not separately controlled. Accordingly, air staging strategies could not be applied during the combustion tests performed in this unit.

This burner is also equipped with an integrated ash removal device, which automatically scrapes away ashes and other combustion residues and moves them into the combustion chamber. The cleaning frequency can be adjusted to the fuel properties. However, cleanings cannot be performed under a continuous operation mode of the system. The biomass feeding must therefore be interrupted and the burner must cool before ashes are removed from the burner grate.

Combustion temperatures were continuously measured with thermocouples in the vicinity to the burner grate at different positions (top, middle and bottom of the burner area) and in each transfer step in the flue gas heat exchanger.

Continuous measurements of emissions of gaseous compounds ($O_2$, CO and NO) were performed with electrochemical sensors (Flue gas analyzer Testo 350XL) located in the exhaust gas directly after the boiler (see Figure 1). Concentration of $NO_X$ was evaluated as nitrogen dioxide ($NO_2$) equivalent according to the European standard EN 303-5:1999 based on the measured NO content in the flue gases [11]. This approximation was needed because of limitation of the measuring equipment for the $NO_2$ quantification. Nevertheless, smaller amounts of $NO_2$ are expected to be present in the flue gases from biomass combustion compared to NO emissions [12-13].

Total particle mass concentration emissions were determined by isokinetic dust filter sampling in the flue gases according to similar procedures as proposed by the standard EN 13284-1 [14]. In addition, particle mass size distributions was determined by using a 13-step low-pressure impactor placed in the same location in the flue gas channel as the total dust sampling (see Figure 1).

## 1.3. Combustion Conditions

A general description of the combustion process used to evaluate environmental behavior of the three selected fuels is summarized in this section. This method resembles the one described in a previous work [10].

Firstly, initial combustion tests were performed to adjust operating conditions to the specific properties of each fuel in order to attain as good as possible results in this unit. Secondly, reproducibility tests were carried out at the selected operating conditions determined by the initial tests. Burner and combustion conditions were established as the ones needed to achieve a heat output as close as possible to the nominal value of the burner. Total amount of air supplied was also regulated to lower the emissions and to attain as complete burnout as possible. The burnout effectiveness was defined based on tests of loss of ignition matter in solid residues and by the unburned CO in flue gases.

Due to the fact that the fuel feeding had to be stopped during cleaning periods, the operation period before cleaning was set to be as longest as possible. During the initial tests with brassica and cassava, some operational problems were observed since the burner showed severe problems to handle the high amount of ash formed in the burner. This caused unwanted operational shut-downs.

In order to enable longer operation periods, the brassica and cassava fuels were therefore blended in the proportion of 50 wt%, wet basis (w.b.) with standard high quality stem-wood based softwood pellets. Ash content for the softwood pellets was extremely low and around 0.35 wt% d.b. Considering these conditions, the ash content was accordingly adjusted to 3.7 wt% d.b. for the blend brassica-woody pellets ($br_{50\%}$-$w_{50\%}$) and around 2.2 wt% d.b. for the blend cassava-woody pellets ($cassG_{50\%}$-$w_{50\%}$). Nevertheless, despite using the blends, the maximum combustion period for brassica-woody pellets was around 30 min before the burner cleaning had to be initiated For poplar, combustion period was rather close to 2 h and for cassava-woody pellets around 1.5 h.

According to the above described procedure applied during the combustion tests in this work, the gaseous and particle emission measurements can be considered to be representative for what may be expected under normal and efficient running conditions using these fuels in this kind of small scale burner system. Data was collected for further calculation of the combustion performance parameters after a steady-state regime was reached at the fixed operating conditions selected for each fuel, and determined by the preliminary experiments.

## 1.4. Efficiency and Emission Threshold Values

Data obtained for emissions and thermal efficiency were compared with the established requirements for combustion of standardized fuels. For this comparison, both the European standard EN 303-5:1999 [11] and the stricter Austrian law art. 15a B-VG 2010 [15] were considered.

The standard EN (European norm) 303-5:1999 is used to test boilers of nominal output < 300 $kW_{th}$ firing standardized woody fuels [11]. This standard defines limits for the thermal efficiency and the emissions of CO, $C_XH_Y$ and particulate matter in terms of three efficiency classes. It also suggests measurement of $NO_X$ emissions although no limits are established for them yet. Emissions limits established by this standard are presented in Table 1.

More restrictive threshold values are defined by the Austrian agreement 15a B-VG:2010 [15-17], which specifies limits related to the thermal efficiency and the CO, $NO_X$ and particulate matter emissions. Values are specific for commercial heating systems of nominal heat output up to 400 $kW_{th}$ firing standardized high quality woody fuels and non-woody biofuels (straw, bark, cereals and mixtures) according to ÖNORM standards. A summary of limiting values for thermal efficiency and emissions established by the Austrian agreement is also included in Table 1.

**Table 1. Emission regulations for automatically loaded combustion system firing solid biofuels**

| Biofuel type | Document | Nominal heat output, $kW_{th}$ (PN) | Particles[a] | | CO [a] | | $NO_X$ | Efficiency, % ($\eta$) | |
|---|---|---|---|---|---|---|---|---|---|
| Standardized woody fuels | European Norm (EN) 303-5:1999 | 0 < PN ≤ 50 | Class 1 | 200 mg/Nm³ | Class 1 | 15000 mg/Nm³ | --- | Class 1 | $\eta$=47+6*log PN |
| | | | Class 2 | 180 mg/Nm³ | Class 2 | 5000 mg/Nm³ | | Class 2 | $\eta$=57+6*log PN |
| | | | Class 3 | 150 mg/Nm³ | Class 3 | 3000 mg/Nm³ | | Class 3 | $\eta$=67+6*log PN |
| | Austrian agreement (15a B-VG) | up to ≤ 400 | 50 mg/MJ[c] | | [b] 500 mg/MJ | | [c] 150 mg/MJ | 90 | |
| Non-woody standardized biofuels | Austrian agreement (15a B-VG) | up to ≤ 400 | 60mg/MJ[c] | | [b] 500 mg/MJ | | 300 mg/MJ | 90 | |

a. Limit related to 10% $O_2$ d.g. at 0°C and 1013 mbar [11].
b. At partial load (30% nominal heat output) limit increase up to 750 mg/MJ [15].
c. Future requirements (1.1.2015) are expected to be set at 100 mg/MJ for $NO_X$ and at 25 mg/MJ for particle emissions from woody sources and at 35 mg/MJ for particle emissions from non-woody sources [12].

## 1.5. Operating Parameters

As mentioned in section 1.3, combustion parameters were adjusted to achieve a reasonable high fuel load, as close as possible to the boiler nominal heat output. Generally, load input in the tests ranged around 20 $kW_{th}$ for the three fuels used. Total lambda was around 2, as expected for this type of burner and combustion technology [9, 12]. These conditions lead to a high combustion efficiency evaluated at the corresponding flue gas temperature, around 90% for poplar and around 93% for brassica and cassava. Comparison of these results with the restrictions set by the standards EN 303-5 for class 3 and the Austrian agreement (15a B-VG) for an automatically loaded 25$kW_{th}$ unit indicated that efficiency satisfied the requirements.

## 1.6. Emissionsperformance of the Fuels

In Table 2, a summary of the environmental behavior of the three fuels with regard to CO, $NO_X$ and particulate matter emissions is presented. Results for CO and $NO_X$ emissions are given as average values with standards deviations. As mentioned previously, the ash characteristics of the tested fuels caused some restrictions on the performance of the combustion tests.

Under these conditions, CO concentration levels from each fuel tested were under the EN 303-5:1999 requirements set for the highest efficiency class of boilers, class 3 (see Tables 1 and 2). With respect to $NO_X$ emissions, the given $NO_X$ concentrations from the measurements might be somewhat lower than the real values because of the non-quantified contribution of $NO_2$ emissions, as indicated in section 1.2. The highest emissions were seen from the herbaceous and the agricultural residue fuels and resulted to be almost more than twice the $NO_X$ emissions determined for poplar (see Table 2). The results for brassica and cassava were not satisfactory enough if they are compared to the Austrian restrictions. According to the expected $NO_X$ formation mechanism for this type of combustion units (i.e., the "fuel-$NO_X$" mechanisms) accounted for in detail elsewhere [12-13], main differences for the level of this pollutant among fuels are basically attributed to the total fuel-N content. As indicated in section 1.1, N content for brassica and cassava was approximately 5 times higher than the one for poplar.

**Table 2. Emissions of CO, NOx and total particulate matter from the tested fuels**

| Parameter | Unit | Poplar | Brassica | Cassava G |
|---|---|---|---|---|
| CO | mg/Nm³(10 % $O_2$ d.g.) | 449±118 | 225±103 | 996± 380 |
| | mg/MJ | 261 ± 62 | 133 ± 47 | 621 ± 200 |
| $NO_X$ | mg/Nm³(10 % $O_2$ d.g.) | 269 ± 23 | 478 ± 86 | 756 ± 277 |
| | mg/MJ | 157 ± 10 | 292 ± 43 | 457 ± 67 |
| Particles | mg/Nm³(10 % $O_2$ d.g.) | 324 | 665 | 623 |
| | mg/MJ | 176 | 342 | 340 |

Furthermore, dilution factor by using a 50% blend certainly affected the $NO_X$ emission results obtained from this work. Accordingly, concentration of $NO_X$ might be higher than reported values in Table 2.

Regarding the particles emissions, the results for the tested fuels were generally significantly higher than the limits defined by the EN 303-5:1999 for boilers class 3, even also if results are compared to the restrictions set for boilers class 1 (see Table 1). Considering specific emissions, results for poplar were according to requirements of the Austrian agreement (15a B-VG). Nevertheless, for cassava and brassica the determined particulate emissions did not satisfy the established specific limits. In all the cases, the particle emissions from this kind of small scale systems and fuels are totally (>95%) dominated by fine (<1 μm) particles, as was determined by the impactor measurements carried out in this work.

# 2. Measurements for Controlling Emissions

The experiments showed that combustion of the fuels was attained under an acceptable level of CO emissions. However, rather high level of $NO_X$ emissions and, considerably high formation of fine particle emissions was also identified for the herbaceous crop (brassica) and for cassava.

Taking into account the current burner unit design and the results obtained from the combustion experiences presented here, further research and development lines can be identified to handle operational problems caused by the fuel ash characteristics and also to improve the environmental performance of these biofuels during fixed bed combustion.

Improvements of the design and operation features of the conversion system might provide a better control on the ash fuel properties leading to a minimization of the maintenance needs and increasing reliability. In addition, to reduce the particle and $NO_X$ emissions incorporation of primary or secondary reduction measures might be needed, of special relevance for medium scale (0.1-10 $MW_{th}$) applications.

Primary measures aim at affecting the combustion process and/or the ash transformation directly occurring in the burner/on the grate, to avoid or modify ash related operational problems and/or formation of pollutants. Secondary measures are, however, applied after the heat exchanging devices with the purpose of removal/reducing the pollutants from the exhaust gas flow. Primary measures seem also to be economically more attractive than secondary measures. Nevertheless, when these primary measures are insufficient, the incorporation of secondary measures is considered to be essential. In some aspects, the potential of using primary processes and fuel related measures (e.g., for particle emission reduction), has not yet been fully explored and evaluated.

As mentioned, air staging strategies to control emissions could not be applied during the combustion tests presented here because of some burner limitations. However, it is well known that this type of primary measures can be useful to fulfill standard regulations for $NO_X$ emissions, and also proven to be promising for particle reduction. Accordingly, restrictions of the system to offer a better control of the air supply, total amount and distribution (primary and secondary air) adapted to fuel quality variations would had certainly affect the results for $NO_X$ and particulate matter emissions.

Small scale biomass combustion is an important source of particulate pollution to ambient air in many European countries and elsewhere all over the world [18-20]. This situation is primarily related to the extensive use of rather old, un-efficient and simple boilers, stoves and fireplaces burning wood and other biomasses. However, concerns should also be taken when introducing new biomass based technology, for instance, as alternative to other heating systems on the residential market. Exposure to particulate pollution in the ambient air is today globally considered as a major public health problem and all risk assessments and air quality guidelines are based on particle mass concentrations.

Thus, it is important to assess and control the particle emissions from different sources, not at least the presently studied biomass combustion technology and fuels. In general, combustion systems generate fine (<1 μm) particle emissions composed by a varying mixture of soot, organics and inorganic ash components [18-20]. A higher combustion efficiency can provide a lower formation of soot and organics (i.e., products of incomplete combustion). In automated small and medium scale biomass combustion systems like, for example, the one used in this work, the combustion conditions can normally be adjusted to achieve high efficiency and low emissions of products of incomplete combustion in the particulate matter. In these cases, the fine particles are composed of ash species vaporized in the fuel conversion, which subsequently condense as fine particles in the flue gases [20]. The formation and emission of these small particles in biomass combustion appliances is very complex and influenced by several factors related to fuel and ash properties, conversion unit design, operation parameters, boiler and flue gas design and potential cleaning devices. A further description of the particle formation mechanisms and physical and chemical characteristics can be found in the relatively vast literature in this field. All these issues would have to be taken into account if this type of fuels was actually considered to participate in the biomass heating market in the close future.

# Conclusion

Environmental behavior of three ash-rich pelletized novel biofuels, two dedicated energy crops (i.e., *Brassica carinata* and *Populus sp.*) and one agricultural residue (i.e., *Manihot esculenta*) was assessed during their combustion in a residential boiler facility with an output of 25 $kW_{th}$. The experimental results from this study revealed that main implication concerning the combustibility of these fuels was related to burner limitations to handle the large amount of ash accumulated in the burner. Nevertheless, by certain adjustments of the burner and blending of pellets during the experimental work performed here, limitations were managed and an evaluation of the environmental performance related to the emissions for the tested fuels was carried out.

Under the selected operating conditions for the fuels, combustion quality could be achieved under acceptable level of CO emissions and efficiency both for the two Spanish energy crops and the Chinese agricultural residue fuel. Concerning the $NO_X$ emissions, only acceptable results were obtained for poplar. For brassica and cassava tests, the results were generally high if they are compared to the present European regulations. Particulate emissions from the three fuels tested exceeded the limiting values defined by the Austrian agreement. The undesirable high levels of these two pollutants ($NO_X$ and particles) were mainly dependent on the inherent fuel properties (i.e., high nitrogen and ash contents). Due to the brassica and cassava conversion showed to be certainly troublesome, blending with high quality woody pellets (low N and ash contents) was used during this project. Though combustion performance improved, still $NO_X$ and fine ash particles emissions were rather high.

Consequently, the incorporation of primary or secondary particle precipitating reduction measures and other initiatives to achieve $NO_X$ reduction might be needed. In addition, due to the high ash content, specially designed burners/grate units would be required if a utilization of these fuels in small and medium scale combustion systems seeks being feasible. Without these modifications, the applicability of introducing this kind of biofuels to the residential heating sector seems to be rather limited.

## Acknowledgments

The present work was financially supported by the collaboration between the Swedish Institute Guest Scholarship Program for Research in Sweden 2008/09, which awarded Maryori Díaz-Ramírez, and Umeå University, which was her host university in Sweden, and the Spanish Education and Science Ministry, through Project "Bio3", ENE2008-03194/ALT, and Project PSE "On Cultivos", PS-120000-2005-6. Thanks are also owed to the Swedish Energy Agency (30646-1, WP7).

## References

[1]  Berndes G., Hansson J. (2007). Bioenergy expansion in the EU: Cost-effective climate change mitigation, employment creation and reduced dependency on imported fuels. *Energy Policy*, 35 (12): 5965-5979.

[2]  Tuck G., Glendining M.J., Smith P., House J.I., Wattenbach M. (2006). The potential distribution of bioenergy crops in Europe under present and future climate. *Biomass Bioenerg.*, 30 (3): 183-197.

[3] Gasol C.M., Martinez S., Rigola M., Rieradevall J., Anton A., Carrasco J., Ciria P., Gabarrell X. (2009). Feasibility assessment of poplar bioenergy systems in the Southern Europe. *Renew. Sust. Energ. Rev.*, 13 (4): 801-812.
[4] Martínez-Lozano S., Gasol C.M., Rigola M., Rieradevall J., Anton A., Carrasco J., Ciria P., Gabarrell X. (2009). Feasibility assessment of Brassica carinata bioenergy systems in Southern Europe. *Renew. Energy*, 34 (12): 2528-2535.
[5] Capablo J., Jensen P.A., Pedersen K.H., Hjuler K., Nikolaisen L., Backman R., Frandsen F. (2009). Ash Properties of Alternative Biomass. *Energy Fuels*, 23: 1965-1976.
[6] Ciria M.P., Solano M.L., González E., Fernández M., Carrasco J.E. Study of the variability in energy and chemical characteristics of brassica carinata biomass and its influence on the behavior of this biomass as a solid fuel, in: $2^{th}$ World Conference and Technology Exhibition on Biomass for Energy, Industry and Climate Protection, Roma, Italy, 2004, pp. 1461-1464.
[7] Staiger B., Unterberger S., Berger R., Hein K.R.G. (2005). Development of an air staging technology to reduce $NO_X$ emissions in grate fired boilers. *Energy*, 30 (8): 1429-1438.
[8] Monti A., Di Virgilio N., Venturi G. (2008). Mineral composition and ash content of six major energy crops. *Biomass Bioenerg.*, 32 (3): 216-223.
[9] Sippula O., Hytönen K., Tissari J., Raunemaa T., Jokiniemi J. (2007). Effect of Wood Fuel on the Emissions from a Top-Feed Pellet Stove. *Energy Fuels*, 21 (2): 1151-1160.
[10] Díaz-Ramírez M., Boman C., Sebastián F., Royo J., Xiong S., Boström D. (2012). Ash Characterization and Transformation Behavior of the Fixed-Bed Combustion of Novel Crops: Poplar, Brassica, and Cassava Fuels. *Energy Fuels* 2012; 26 (6): 3218-3229.
[11] European standard EN 303-5:1999. Heating boilers for solid fuels, hand and automatically stoked, nominal heat output of up to 300 kW. *European Committee for Standardization* (CEN). 1999.
[12] Obernberger I., Brunner T., Bärnthaler G. (2006). Chemical properties of solid biofuels--significance and impact. *Biomass Bioenerg*, 30 (11): 973-982.
[13] Johansson L.S., Leckner B., Gustavsson L., Cooper D., Tullin C., Potter A. (2004). Emission characteristics of modern and old-type residential

boilers fired with wood logs and wood pellets. *Atmos. Environ*, 38 (25): 4183-4195.
[14] European standard EN 13284-1:2001: Stationary source emissions - Determination of low range mass concentration of dust- Part 1: Manual gravimetric method. 2001.
[15] Art. 15 a B-VG agreement: Precautionary measures regarding small-scale heating systems with a nominal heat output up to 400 kW for residential heating. 2010.
[16] Zeng T. National conditions-Austria. IEE/09/758/SI2.558286 - *MixBioPells*. WP4 / D 4.3, 2011.
[17] Zeng T., Pollex A. Overview about the national conditions. IEE/09/758/SI2.558286 - MixBioPells. WP4 / D 4.3, 2011.
[18] Nussbaumer T., Czasch C., Klippel N., Johansson L.S., Tullin C. Particulate emissions from biomass combustion in IEA countries. 2008.
[19] Boman BC, Forsberg AB, Järvholm BG. (2003) Adverse health effects from ambient air pollution in relation to residential wood combustion in modern society. Scand Work, *Environ Health*, 29:251-260.
[20] Kocbach Bølling A, Pagels J, Yttri K-E, Barregard L, Sällsten G, Schwarze PE, Boman C. (2009). Health effects of residential wood smoke particles: the importance of combustion conditions and physicochemical particle properties. *Part. Fibre Toxicol.*, 6:29.

# Index

## A

abiotic factors, vii, x, 113, 115, 118, 123, 124, 127
access, 12
accounting, 41, 44
acetylcholine, 38
acid, 40, 42, 43, 51, 63, 64, 98, 103, 104
active transport, 46, 48, 56
adaptability, 3
adaptation(s), 8, 31
ADC, 58
additives, 103
adenine, 42
adjustment, vii, 2, 14, 29
adults, 8, 9, 10, 38
adverse effects, 11, 97
Africa, ix, 70, 74, 83, 114, 115, 129
age, 118
agriculture, 131
air pollutants, 109
air quality, 143
alkaloids, vii, viii, 13, 37, 38, 40, 41, 42, 43, 45, 46, 47, 48, 49, 52, 56, 57, 59, 61, 63
allele, 81, 85
allopolyploid, 61, 90
ALT, 145
ambient air, 110, 143, 147
amino acid(s), 41, 45, 58, 64, 98
ammonium, 103

amplitude, 11
aneuploid, 74
animal behavior, 13
anthocyanin, 52, 62, 65
apex, 46, 74
Arabidopsis thaliana, 65, 85
arginine, 41
aspartate, 40, 42
aspartic acid, 98
assessment, 31, 115, 129, 146
atmosphere, 111
Australasia, ix, 69
Austria, 147
autoimmunity, 86
avoidance, 4
awareness, 25

## B

barriers, 70, 71, 72, 73, 87
base, 17, 79
beams, 89
Belgium, 113, 129
benefits, 12
bioaccumulation, 96
bioavailability, 98
biochemistry, 57
biodiversity, vii, 2, 3, 4, 12
bioenergy, 145, 146
biofuel, 135

biological control, 8
biological processes, 28
biological systems, 18, 96
biomarkers, 56
biomass, x, 4, 6, 8, 25, 35, 105, 107, 111, 114, 115, 120, 127, 128, 129, 134, 135, 138, 143, 146, 147
biosynthesis, vii, 43, 45, 48, 49, 51, 52, 53, 54, 55, 56, 57, 58, 59, 60, 61, 62, 63, 65, 66
biosynthetic pathways, 41
biotechnology, 57
biotic, 4, 11, 128
biotic factor, 8
blends, 139
blood, 38
bloodstream, 38
boilers, 139, 141, 142, 143, 146, 147
brain, 38
brassica, vii, 135, 136, 139, 141, 142, 144, 146
Brazil, 1, 38
breeding, 44, 70, 73
browsing, vii, 2, 3, 8, 9, 10, 13, 33, 34
burnout, 139

## C

Cameroon, 113, 116, 127, 129, 130, 132
cancer, 111
carbohydrate, 43
carbon, 136
carboxylic acid, 55
carcinogen, 44, 61
cassava, vii, 135, 136, 139, 141, 142, 144
Catharanthus roseus, 54
cation, 42, 43, 120
cattle, 13, 31, 33
cDNA, 42, 48, 50, 58
CEC, 120
cell biology, 57
cell culture, 41, 55, 58, 60
cell death, 85, 93, 94
cell division, 71
Central African Republic, 130

central nervous system, 38
certification, 137
challenges, 3, 12
chemical(s), x, 11, 13, 23, 38, 41, 44, 47, 50, 103, 104, 109, 114, 127, 134, 144, 146
chemical characteristics, 127, 144, 146
chemical properties, 134
chemical reactions, 114
chimpanzee, 130, 131
China, x, 29, 134, 135
chiral center, 41
chloroplast, 75
chromatography, 104
chromosome, ix, 70, 72, 76, 77, 78, 79, 80, 81, 82, 85, 87, 89, 90, 93, 94
cities, 110, 111
classes, 117, 140
classification, 89
cleaning, 109, 138, 139, 144
climate, 116
climate change, 129, 145
clone, 85
cloning, 41, 58, 62
cocaine, 38
coding, 16, 18
collaboration, 145
colonization, 30
color, 74, 79, 118
combined effect, 115
combustibility, 136, 144
combustion, x, xi, 134, 135, 136, 137, 138, 139, 140, 141, 142, 143, 144, 145, 147
commercial, 49, 135, 140
community(ies), vii, x, 2, 3, 4, 5, 6, 11, 12, 23, 25, 27, 30, 34, 35, 113, 114, 115, 117, 118, 120, 123, 127, 128, 129, 130, 131, 132
compensation, 14
competition, 4, 5, 11, 12, 132
complementarity, 4, 14
complex interactions, 4
complexity, 3, 25
compliance, 108

# Index

composition, 6, 11, 23, 88, 103, 104, 115, 117, 146
compounds, 47, 48, 56, 57, 97, 138
condensation, 43
conductance, 110
configuration, 41
Congo, 116
Congress, 29, 35, 90
conjugation, 51
consensus, 54
conservation, 3, 6, 31, 54
conserving, 3
consumers, 2
consumption, vii, 2, 8, 10, 16, 19, 24, 27, 38
contaminated soil, 111
contaminated water, 96
contamination, 96, 97, 100, 108, 109, 110, 111
coordination, 46
copper, vii, ix, 42, 95, 99, 105, 109
correlation(s), 18, 118, 120, 123, 125, 127, 128
correlation coefficient, 18
cortex, 46
cotton, 98
cotyledon, 74, 79
covering, 104, 106
critical period, 6
Croatia, 94
crop(s), x, 98, 102, 103, 134, 135, 142, 144, 145, 146
cultivars, viii, 37, 86
cultivation, ix, 49, 55, 74, 96, 103
culture, 73, 74, 87, 88, 90, 92, 93
culture medium, 49, 55, 72
current limit, xi, 134
cycles, 32
cytochrome, 40, 44, 60, 61
cytokinins, 49, 88
cytology, 89

## D

data collection, 129
data processing, 109
data set, 82
database, 58
decomposition, 127
defence, 87
defense mechanisms, 59
defensive toxins, viii, 37, 56
deficiency, 96, 107
degradation, 51, 108
demographic structure, 9
demography, 5, 7, 8, 9, 27, 131
dependent variable, 118, 120
deposition, ix, 11, 95, 99, 108, 109
depression, 101, 102, 106, 107
depth, 104, 107, 117
derivatives, 41, 42, 44, 50, 94
detoxification, 62, 98
diamines, 42
diet, viii, 2, 14, 16, 17, 24, 35, 130, 131
diet composition, 3
diffusion, 47
digestibility, 3, 16, 17, 20, 22, 33
digestion, 13, 32
dimerization, 52
diploid, 71, 87
direct observation, 16, 18, 29
diseases, 86
dispersion, 104
distilled water, 103
distribution, 7, 8, 9, 15, 85, 103, 132, 136, 137, 143, 145
divergence, 83
diversification, 42
diversity, vii, 2, 3, 5, 7, 8, 11, 14, 17, 18, 25, 29, 34, 89, 114, 115, 120, 127, 131
DNA, 66, 80, 87, 91, 93
dominance, 7, 8, 55
Dominican Republic, 131
dopamine, 38
dosage, 98
down-regulation, 49, 56
drought, 12, 131, 132
dry matter, 15, 17, 19, 20
dwarf stems, x, 114, 118, 127, 128

## E

ecological processes, 3, 4
ecology, 7, 20, 34, 35, 50, 56, 130, 131, 132
ecosystem, 4, 25, 31, 96, 100, 111, 131
egg, 73, 92
elucidation, 49
emission, x, 100, 134, 139, 142, 143, 144
employment, 145
encoding, 45, 58, 79
endosperm, 72
energy, x, 13, 16, 24, 35, 134, 135, 144, 146
engineering, 57, 58, 60, 66
enlargement, 14
environment(s), 4, 11, 13, 14, 19, 34, 35, 97, 110, 111, 130
environmental change, 11, 31, 35
environmental characteristics, 115
environmental conditions, x, 5, 114, 127
environmental degradation, 105
environmental factors, 115
environmental stress, 98
environmental variables, x, 114, 118, 119, 120, 127
enzymatic activity, 43
enzyme(s), viii, 37, 40, 43, 44, 45, 51, 54, 58, 59, 64, 98
epidermis, 98
Equatorial Guinea, 116
equilibrium, 4
equipment, 138
erosion, 26
EST, 41
ethylene, viii, 37, 55, 66, 67
EU, 145
euphoria, 38
Europe, 34, 145, 146
evaporation, 102
evidence, 59, 82, 92, 93, 130
evolution, 29, 32, 54, 61, 81, 83, 87, 90, 91, 92, 129
extraction, 103
extrusion, 48, 61, 62

## F

FAD, 43
families, x, 43, 48, 113, 116, 125, 127
family factors, 52
family members, 43, 51, 54
farmers, 3, 6, 13, 23, 25
fertility, x, 94, 114, 115, 120, 127, 132
fertilization, 72, 87, 98, 107
filtration, 104
flavonoids, 48
flavor, 13
flexibility, 6
flowering period, 6
flowers, 14, 71, 118, 125, 128
flue gas, 138, 139, 141, 144
fluorine, 96
food, 3, 13, 16, 19, 20, 23, 33, 34, 131, 132
food chain, 97
food intake, vii, 2, 32
forbs, 19, 22
forest ecosystem, 111
formation, xi, 42, 43, 44, 50, 51, 52, 54, 59, 61, 65, 70, 74, 100, 110, 134, 141, 142, 143, 144
France, viii, 1, 2, 14, 27, 29, 33
free radicals, 98
freezing, x, 96, 105, 109
fruits, 14, 118, 125, 128
fusion, 49, 72, 89, 91

## G

Gabon, 130, 131, 132
gene expression, 55, 66
gene promoter, 53
genes, viii, ix, 37, 41, 43, 44, 46, 48, 49, 50, 51, 52, 53, 54, 56, 57, 58, 59, 60, 61, 62, 65, 66, 69, 73, 75, 78, 79, 80, 81, 82, 85, 92, 93
genetic background, 79
genetics, 62, 81, 89, 91
genome, 41, 50, 61, 88, 90, 91
genomics, 59

genus, 38, 41, 70, 74, 86, 87, 92, 93, 94
germination, 11, 73, 79, 102, 103, 127, 128
glutamic acid, 98
glutamine, 75, 87
glycoside, 43
GPS, 117
grass(es), ix, 3, 5, 6, 11, 12, 18, 19, 21, 22, 30, 31, 32, 34, 96, 100, 101, 102, 104, 106, 107, 108, 109, 110
grasslands, 3, 11, 29, 32, 34
gravity, 21
grazers, 13
grazing, vii, 2, 3, 4, 6, 7, 8, 9, 10, 11, 12, 13, 14, 18, 19, 22, 23, 24, 25, 26, 27, 28, 29, 30, 31, 32, 33, 34, 35
greenhouse, 102, 103
greening, 100
growth, 5, 6, 7, 8, 9, 11, 25, 27, 55, 58, 63, 64, 71, 73, 79, 92, 93, 96, 98, 99, 106, 110, 111, 115, 127, 129, 131, 132
growth dynamics, 6
growth rate, 8, 25, 98
guidelines, 12, 25, 143

## H

habitat(s), x, 29, 38, 110, 113, 115, 116, 117, 118, 119, 120, 121, 123, 125, 127, 128, 129, 130, 132
habitat quality, 115
half-life, 18
haploid, 71, 80
harvesting, 55
haze, 110
health, 97, 107, 108, 109, 147
health effects, 147
heavy metals, 96, 97, 98, 99
height, 9, 12, 102, 104, 106, 116, 118
helium, 89
herbaceous-shrub mosaics, vii, 2
herbivores, vii, 2, 3, 4, 8, 12, 13, 14, 23, 31, 33, 35, 47, 55, 65, 115
heroin, 38
heterogeneity, vii, 2, 3, 4, 7, 11, 12, 29, 33, 34

history, 49, 56, 115, 130
horses, 27
host, 66, 145
House, 145
human, 38, 97, 109, 115
human health, 38
humidity, 99, 117, 118, 120, 124, 125
hybrid, vii, viii, 52, 69, 70, 71, 72, 73, 74, 75, 76, 77, 78, 79, 80, 81, 82, 83, 84, 85, 86, 88, 90, 91, 93, 94
hybridization, ix, 69, 70, 72, 78, 86, 87, 91, 92, 94
hydroponics, ix, 96, 102
hypocotyl, 74
hypothesis, 20, 75

## I

identification, 49, 56, 58, 88, 93
identity, 50
idiosyncratic, 127
IEA, 147
immunolocalization, 45
improvements, xi, 134
in vitro, 73, 87, 91, 92, 93
in vivo, 33
incompatibility, 71, 72, 86, 90, 91, 92, 93
incomplete combustion, 144
incongruity, 71, 89
individuality, 13
individuals, 25, 44
induction, 52, 55, 58, 61, 71
industrial emissions, 96, 99, 109
industry, 38, 110
infection, 85
ingestion, 12, 13, 19, 20, 21, 22, 31, 33, 35
inheritance, 60
initiation, 52, 65
insecticide, 38
insects, viii, 8, 32, 37, 50
integration, vii, 2
interface, 53
interference, 43, 49, 52, 54, 60
intervention, 100
investments, 135

irradiation, 72, 73, 92
isoflavone, 43, 59, 60
isolation, 49, 56, 70, 81, 82, 83, 91, 92
isoleucine, 51, 64
issues, 96, 144
Italy, 146

## J

Japan, 37, 57, 69, 90, 94
juveniles, 8, 10, 25

## K

kinetics, 18, 32
Kola Peninsula, ix, 95, 99, 100, 111

## L

landscape, vii, 1, 23
lead, viii, 2, 59, 93, 96, 134, 136, 141
learning, 12, 13
Lepidoptera, 66
lesions, 85, 92
life cycle, 28
light, x, 11, 12, 31, 115, 117, 120, 125, 127, 129
light conditions, 114
livestock, 2, 3, 4, 11, 13, 25, 27, 29, 30, 34
localization, 43, 45, 48
loci, viii, 37, 43, 48, 49, 62, 63
locus, viii, 38, 45, 50, 52, 53, 54, 55, 66, 80, 81, 85
logging, 131
lysine, 40, 45

## M

magnitude, 115, 127
majority, 19, 74, 115
management, vii, 2, 3, 4, 6, 10, 12, 13, 14, 24, 25, 27, 28, 29, 31, 32, 33
manipulation, 56

mass, 3, 11, 12, 14, 15, 16, 18, 19, 21, 22, 29, 32, 55, 102, 104, 138, 147
materials, ix, 41, 96
matrix, 104, 118
matter, x, 120, 121, 134, 135, 136, 139, 140, 141, 142, 143, 144
measurement(s), 138, 139, 140, 141, 142
media, 88
Mediterranean, 5, 30, 32, 34, 35
mentor, 72
mesophyll, 98
metabolic pathways, 49, 54
metabolism, 49, 57, 61, 62, 98
metabolites, 13, 35, 38, 44, 48, 56, 57, 63, 64
metal ion(s), 48
metallurgy, 110
metals, ix, 95, 96, 97, 98, 103, 104, 105, 108
methylation, 42
micronutrients, 97
microscopy, 48
migration, 111
modelling, 31
models, 7, 8, 13, 17, 18, 20, 21, 30, 35
modern society, 147
modifications, 145
moisture, 12
molecular cytogenetics, 88
molecules, 42, 63
morphology, 11, 89, 90
mortality, 127, 128
mosaic, 85, 87, 88, 94, 120
motif, 51, 64, 65
mRNA, 49
multidimensional, 118
mutagenesis, 44
mutant, viii, 37, 42, 43, 49, 50, 54, 56
mutation(s), 44, 45, 61

## N

$Na^+$, 48, 105
NAD, 40, 42, 54, 59
Namibia, 74

narcotics, 38
native population, 63
natural habitats, 115
necrosis, 65, 73, 86, 88
negative effects, 97
nervous system, 38
Netherlands, 56, 57, 66
neurotransmitters, 38
neutral, 34
New Zealand, 30
nickel, vii, ix, 95, 99, 105, 109
nicotiana, vii, 94
nicotinamide, 42
nicotine, viii, 37, 38, 40, 41, 42, 43, 44, 45, 46, 47, 48, 49, 50, 51, 52, 53, 54, 55, 56, 57, 58, 59, 60, 61, 62, 63, 64, 65, 66
nicotinic acid, 40, 42, 43, 45, 59
nitrogen, 11, 34, 114, 117, 120, 121, 129, 138, 145
nitrogen dioxide, 138
nitrosamines, 61
nodes, 82
nodules, 11
nucellus, 71
nuclear genome, 78
nutrient(s), x, 3, 6, 11, 13, 14, 17, 96, 108, 109, 114, 117, 127, 129
nutrition, 72
nutritional status, ix, 96, 102

# O

obstacles, 70
oil, 96
operating costs, 135
opportunities, 27
optimization, 20, 21, 42
organ(s), 8, 10, 14, 19, 23, 25, 27, 47, 48
organelle, 43, 47
organic compounds, 38
organic matter, 16, 17, 22, 117, 120
ornithine, 40, 41, 58
overlap, 118
overproduction, 63
ovule, 72, 87, 88, 90, 92

oxidability, 104
oxidation, 45
oxygen, 43
ozone, 110

# P

Pacific, 83
paclitaxel, 63
paradigm shift, 29
parenchyma, 46
parents, 73
particle mass, 138, 143
pasture(s), vii, 1, 3, 6, 12, 13, 14, 17, 20, 27, 28, 32, 35
pathways, 63
PCA, 118, 119, 120
peat, 99, 102, 107
permission, 129
pH, 48, 99, 103, 117, 120, 121
phenotype(s), 44, 47, 49, 50, 74, 76, 77, 82
phloem, 46, 50
phosphate, 41, 43, 58
phosphorous, 103, 104
phosphorus, 114, 117, 120, 121, 129
photosynthesis, 110, 114
phylogenetic tree, 82, 84
pistil, 71, 89
plant growth, 73, 97, 128, 129
plasma membrane, 47, 48, 49
plastid, 59, 75, 82, 87
pollen, 71, 73, 89, 92, 93
pollen tube, 71, 89, 93
pollination, 71, 72, 87, 90, 93
pollutants, ix, 95, 97, 99, 100, 103, 109, 110, 111, 143, 145
polluters, 110
pollution, ix, x, 95, 96, 97, 98, 99, 101, 103, 105, 106, 109, 110, 111, 134, 143, 147
polyamines, 42
polymorphism, 48
polypeptide(s), 52, 88
polyploid, 74, 90
pools, 42, 63
population, 5, 7, 8, 9, 10, 25, 28, 30, 31, 33

population growth, 7, 8
population structure, 8, 9
porosity, 102
Portugal, 38, 95
positive correlation, 114
positive feedback, 23
potassium, 117
precipitation, ix, 96, 103, 105
principal component analysis, 118, 119, 120
principles, 137
producers, 2
programming, 25
project, 99, 105, 109, 145
promoter, 45, 52, 54, 55
propagation, 117
proteasome, 51, 52, 53
protection, 103
proteins, 45, 46, 51, 52, 59, 62, 64, 65, 87
public health, 143
purines, 49
pyrophosphate, 43

## Q

quantification, 103, 138
quantitative estimation, 7
Queensland, 35
quinolinic acid, 43

## R

race, 88
radiation, 83, 93
radicals, 98
rain forest, x, 114, 131
rainfall, x, 114, 116, 118, 124, 125, 128, 129
rangeland(s), vii, 1, 2, 3, 4, 5, 7, 13, 14, 19, 23, 29, 31, 32, 33, 34
rape, 99
rape seed, 99
reactions, 40, 42, 43, 87, 114
reallocation of resources, 9
receptors, 38

reciprocal cross, 71, 72, 75, 83, 94
recognition, 3, 25
recovery, 110, 131
regrowth, 23
regulations, 140, 143, 145
rehabilitation, 100
relaxation, 38
relevance, 11, 143
reliability, xi, 134, 143
repellent, 50
repressor, 51, 52, 53, 64, 65
reproduction, 7, 98, 110
reproductive age, 9
requirements, 3, 6, 139, 140, 141, 142
researchers, 70, 73, 97
residues, 45, 135, 138, 139
resilience, 3
resistance, 4, 55, 64, 65, 72, 75, 85, 86, 88, 92, 94, 97, 98, 109, 131
resource availability, 6, 8, 130
resources, 2, 4, 5, 6, 12, 14, 18, 55, 114, 115, 129
response, viii, 3, 9, 10, 11, 23, 33, 34, 35, 37, 43, 50, 51, 52, 53, 54, 55, 85, 86, 87, 115, 131
responsiveness, 49
restrictions, 135, 141, 142, 143
rings, 41, 43, 45
risk, ix, 95, 96, 97, 143
risk assessment, ix, 95, 96, 143
RNA, 43, 49, 52, 54, 60
RNAi, 60
room temperature, 103, 104
root(s), viii, ix, 11, 37, 41, 42, 43, 46, 47, 48, 49, 50, 52, 54, 55, 56, 58, 61, 63, 74, 79, 86, 96, 98, 104, 107, 109
root growth, 98
root hair, 46
root rot, 86
root system, ix, 96, 107
routes, 40
Russia, ix, 95, 99, 111

# S

salinity, 110
saturation, 102
scale system, 142
scaling, 4, 118
science, vii, 2, 20, 34
scope, 136
secondary metabolism, 59, 60, 62
seed, 7, 8, 10, 11, 32, 62, 71, 72, 92, 93, 102, 103
seeding, 101, 102
seedlings, 8, 10, 11, 12, 71, 72, 73, 75, 79, 81, 85, 98, 99
segregation, 81
senescence, 44, 60
sensitivity, 8, 85
sensors, 138
sensory experience, 13
sequencing, 41
services, 31
shade, 128
shape, 114
sheep, 13, 19, 29, 30, 32, 34
shoot(s), 47, 55, 61, 63, 74, 125
shortage, 5, 128
showing, viii, 2, 9, 79, 84, 123
shrub encroachment, vii, 2, 12, 32
shrubland, 16, 30, 35
shrubs, viii, 2, 3, 5, 6, 10, 11, 12, 17, 19, 23, 27, 30, 32, 34, 35
signal transduction, 63
signaling pathway, 60
signalling, 64, 65
signals, viii, 37, 56
signs, 108, 118
simulation, 10
smoking, 38
smoking cessation, 38
$SO_4^{2-}$, ix, 96, 105, 106
social interactions, 33
social learning, 13, 32
social organization, 130
solution, 104, 120
South America, 83
South Pacific, 82, 90
Soviet Union, 110
sowing, 101, 102, 106, 107, 108
Spain, x, 30, 133, 134, 135
spatial learning, 13
speciation, 74, 86, 91
species richness, 3, 131
spectrophotometry, 104
sprouting, 102
SS, 78
St. Petersburg, 110, 111
stability, 14, 25, 26
stabilization, 14, 21, 22
stamens, 87
standard deviation, 104, 121, 127
state(s), 3, 9, 23, 24, 25, 27, 28, 31, 55, 104, 108, 139
statistics, 118
sterile, 71, 88, 89, 91
storage, 46, 47
stoves, 143
stress, x, 62, 63, 96, 98, 110, 111, 120, 129
stress response, 62, 63
structural changes, 14
structural gene, viii, 37, 41, 48, 49, 55, 56
structure, vii, x, 3, 4, 13, 14, 18, 23, 30, 41, 113, 114, 115, 117, 118, 120, 123, 127, 130, 131
structuring, 8, 132
style, 71
substitutions, 44, 81
substrate(s), 42, 44
subtraction, 42, 50
succession, 28, 100, 132
sucrose, 87
sulfur, 96, 103
sulfur dioxide, 96
Sun, 58, 72, 92
supervision, 109
suppression, 43, 44, 45, 49, 51, 52, 60
survival, 7, 11, 25, 58, 108
survival rate, 7
Sweden, 133, 145
symptoms, 73, 74, 82
syndrome, 86

synthesis, viii, 31, 37, 54, 55, 59, 63, 120

## T

target, viii, 24, 25, 27, 38, 51, 52, 54, 98
target organs, 98
taxa, 81, 83, 86
taxation, 38
techniques, 91, 92, 104
technology, xi, 56, 57, 134, 141, 143, 144, 146
temperature, 11, 73, 85, 101, 118, 124, 125, 128, 141
territorial, 1
texture, x, 114, 119, 127
time series, 18, 20
TIR, 85
tissue, 16, 24, 43, 49, 50, 55, 56, 72, 93
tobacco, vii, viii, 37, 38, 41, 42, 44, 45, 46, 47, 48, 49, 50, 52, 54, 55, 56, 57, 58, 59, 60, 61, 62, 65, 66, 73, 85, 86, 87, 88, 89, 90, 91, 93, 94
tobacco smoking, 38
Tonga, 74
tourism, 3
toxic effect, 97
toxic metals, 96
toxicity, 14, 107
toxin, 13
trace elements, 96
tracks, 89
trafficking, 57
traits, 8, 30, 31, 89, 131
trajectory, 28
transcription, viii, 37, 50, 51, 52, 53, 54, 56, 58, 59, 65, 66
transcription factors, 51, 52, 53, 54, 56, 58, 65, 66
transcripts, 43, 45, 55, 66
transducer, 63
transformation(s), 136, 143
translocation, 47, 50, 61
transpiration, 47
transport, viii, 37, 46, 48, 50, 51, 53, 56, 57, 61, 62, 98

treatment, 43, 55, 72, 74
triggers, 38
turnover, 90, 111

## U

ubiquitin, 51, 52
UK, 34, 35, 56, 57
uniform, 3, 5
universe, 57
urban, 131
USA, 26, 30, 35, 58, 59, 60, 62, 63

## V

vacuole, 43, 47
variables, 20, 114, 115, 118, 119, 120, 124, 125, 127, 129, 130
variations, 3, 8, 14, 23, 44, 114, 115, 124, 127, 143
varieties, 49, 135
vasculature, 47
vegetation, vii, ix, 2, 3, 4, 5, 6, 7, 11, 12, 13, 14, 16, 18, 19, 23, 24, 25, 27, 28, 29, 31, 95, 99, 100, 105, 106, 107, 108, 109, 111, 116, 129, 132
vegetative cover, 104
vessels, 46
viruses, 92

## W

Washington, 130
waste, 47
water, 11, 96, 97, 99, 102, 114, 115, 128, 137
weakness, 73
weight gain, 13, 32
weight ratio, 107
wild type, 42
wood, 137, 139, 143, 147
woodland, 32

## X

xylem, viii, 37, 41, 46, 47

## Y

yeast, 48, 49, 52
yield, ix, 70